T0254583

Springer Undergraduate Mathematics Series

More information about this series at http://www.springer.com/series/3423

Daniel Grieser

Exploring Mathematics

Problem-Solving and Proof

Daniel Grieser
Institut für Mathematik
Carl von Ossietzky Universität Oldenburg
Oldenburg
Germany

ISSN 1615-2085 ISSN 2197-4144 (electronic)
Springer Undergraduate Mathematics Series
ISBN 978-3-319-90319-4 ISBN 978-3-319-90321-7 (eBook)
https://doi.org/10.1007/978-3-319-90321-7

Library of Congress Control Number: 2018939472

Mathematics Subject Classification (2010): 00-01, 00A07, 00A09, 97D50

Translation from the German language edition: *Mathematisches Problemlösen und Beweisen* by Daniel Grieser, © Springer Fachmedien Wiesbaden 2013. All Rights Reserved.

Printed on acid-free paper

This Springer imprint is published by the registered company Springer International Publishing AG part of Springer Nature
The registered company address is: Gewerbestrasse 11, 6330 Cham, Switzerland

For Ricarda and Leonard

Preface to the English edition

The classical curriculum for university mathematics education has emphasized calculation and – on a higher level – proofs and systematic development of mathematical theories. In recent years there has been a growing interest in supplementing this with a problem-solving approach. It has become clear that this is not only fun but also very useful as preparation for understanding the mathematical theories, these tremendous advances that mathematics has made over the centuries. These new developments have occurred in Britain, the United States, Germany and certainly other countries as well.

In 2011, the University of Oldenburg introduced a new course into the curriculum: 'Mathematical problem-solving and proving' (Mathematisches Problemlösen und Beweisen). It was aimed at first-year students who have just finished Gymnasium (roughly equivalent to A-levels in England and high school plus epsilon in the US) and are starting a degree in mathematics or mathematics education. This course has been very successful, both in bridging the gap between high school and university, and in providing a fun introduction to the higher mathematics taught at university. Similar courses have sprung up at other universities in Germany, and in Britain and the US, and at other places people are discussing the possibility of introducing them, or at least introducing elements of such an approach into existing courses.

This book arose from my teaching the course twice in Oldenburg, and since many colleagues in English-speaking countries have expressed a great interest in it I decided to translate it into English.

There are many other excellent books on problem-solving. What sets this book apart is that it starts at a very elementary level, but then quite explicitly tries to be a bridge to higher mathematics: the emphasis is on your discovery of mathematics by solving problems, but along the way concepts, notation and terminology of higher mathematics (e.g. sets and mappings) are introduced. In addition, most chapters have a section with additional material (titled *Going further*) which provides a preview of where similar ideas are used in advanced mathematical disciplines.

Contents

Introduction

Tell me and I will forget.
Show me and I will remember.
Involve me and I will understand.
(Lao Tse)

With this book I invite you to come on a voyage of discovery. You will discover mathematics from a completely new perspective: not as a collection of formulas and rules, but as a world which you can explore for yourself, in which you can develop your own ideas and unearth hidden treasures. Don't worry: you are well equipped for your journey, and along the way you will add new tools to your equipment that will help you master difficult stages. I will be your guide. You can decide for yourself whether you want to travel alone, accept a few hints, or let yourself be comfortably guided through the terrain.

At the end you will return from your voyage enriched. Not only will you have gained a new view of mathematics, but you will also take home a wealth of experiences.

Three themes will recur throughout our voyage:

1. Mathematical **problems** and **problem-solving strategies**
2. Mathematical **proofs:** why do we need them, how do we find them, how do we write them?
3. **General ideas** of mathematics

Problems and problem-solving strategies

Problems are the soul of mathematics.
The problems of how to calculate areas and predict the movement of objects led to the invention of the calculus in the 17th century. FERMAT's problem, whether the equation $x^n + y^n = z^n$ has positive integer solutions for $n > 2$, has stimulated number theory up to the

present day. The four colour problem, whether four colours suffice to colour every conceivable map assuming that neighbouring countries always get different colours, led to a new mathematical theory in the 19th century: graph theory. These are big mathematical problems. But there are also countless smaller interesting problems which are accessible for everyone, which you can explore for yourself and thus experience what mathematics is about. You will find such problems in this book.

Problem-solving is fun and creative.
Initially you are in the dark. You look at examples, make a sketch, observe patterns, and by and by you realise what matters and what doesn't. Slowly the darkness lifts, you feel your way forward, develop ideas, and suddenly: Eureka – I've found it! Understanding is deeply satisfying. Being creative makes you happy.

Problem-solving can be learned.
During your voyage you will learn many problem-solving strategies. Some of them are so general that they can also be used outside of mathematics. Others are specific to mathematics. With every problem that you solve you develop your creativity and enlarge your pool of experience.

Problem-solving makes you curious about mathematical theories.
Theories give answers. You will really appreciate an answer only if you have asked the question beforehand, and even more if you have tried to answer it yourself and experienced difficulties. If you learn new theories (like algebra or analysis) in this way, then you will be able to use them well, and advance beyond them.

Proofs

Proofs are the heart of mathematics.
Mathematical proof is quite extraordinary: what is proved today is true – today, tomorrow and in a thousand years. This distinguishes mathematics from all other sciences. Many scientific theories were considered valid for centuries but then had to be corrected. Mathematics also advances, new connections of hitherto unconnected worlds are discovered all the time, problems are solved which had seemed intractable for years; but what is proved once will always be true.

Proofs tame infinity.
By experimenting you may discover that you can find more and more prime numbers, no matter how many you have found already. But does this continue for ever? Are there infinitely many primes? Only a proof can give certainty.

Proofs give certainty.
Many a supposed fact turns out to be a fallacy – and if you have experienced this then you will appreciate what proofs do for you. Without a proof you can never be sure.[1]

Proofs help you to understand.
When you have thought in detail about why a mathematical assertion is true, when you have removed any doubt by logical arguments, then you will understand the assertion itself better, and also you will be better at explaining it to others. And you will remember it better.

Finding proofs is problem-solving and therefore creative. And it can be learned.
A proof is a logically complete argument. But how do we find a proof? Here we need to be creative. Finding a proof is to mastering logic as drawing a picture is to knowing about colours, or as composing a symphony is to knowing about musical notes and harmonies. And just like a picture or a symphony, a proof can be beautiful. There is no general recipe for finding proofs, but just as for general problem-solving there are recurring patterns and ideas. Getting to know these will help you get better at finding proofs. You will find many of these patterns in this book.

General ideas of mathematics

Often mathematics is divided into subdisciplines: geometry, algebra, analysis and so on. However, many ideas occur in all disciplines. In this book you will get to know such ideas by simple examples. You will also learn about general methods of scientific work. You may be surprised to notice that many of them will be familiar to you. We merely name them so you can use them systematically.

[1] See Exercises E 1.10 and E 5.24 and the beginning of Chapter 7.2 for good examples of this.

Among the scientific methods are the cycle of exploring, making hypotheses (also called conjectures in mathematics) and investigating them systematically (in mathematics: proof or disproof); introducing concepts and notation; and identifying essential aspects of a problem and neglecting inessential ones (abstraction).

Among the general mathematical ideas are the extremal principle, the invariance principle and the principle of counting in two ways. You will learn about these ideas here as problem-solving strategies and use them to find out surprising things. But when you delve deeper into mathematics then you will meet these ideas again and again in different guises. They will guide your way in the vast landscape of mathematics.

Who is this book for?

This book is for everyone who likes mathematics and who likes to think about problems that cannot be solved by simply applying a given method. For all who want to improve their problem-solving skills. For mathematically interested high school or secondary school students. For students in colleges and universities who would like to get a different view of mathematics, supplementing their standard maths courses. For teachers in high or secondary schools, colleges and universities who want to teach problem-solving in a more systematic way than just by giving homework problems, maybe even to offer a course that is method-oriented rather than topic-oriented. For leaders of maths clubs or math circles.

As *prerequisites* you only need to know a bit about numbers, basic geometry, how to rearrange equations. You learn these ideas quite early on at high or secondary school. But even if you know some higher mathematics already you will get your money's worth: at many places in the book you will find material on how the elementary ideas introduced here are used in higher mathematics, including present-day research.

Contents

The core of the book is the problems. Lots of problems, their investigation and solution. In the investigation we approach each problem step by step: What are we looking for? What could be a route of approach? Not every route that we try will lead to a solution; sometimes we need to backtrack and start again. Here you experience how mathematics is born, invented, developed. While discovering mathematics you learn how mathematicians think. Here you learn problem-solving strategies. These are collected in the *Toolbox* sections. In Appendix A you will find an overview of all the strategies introduced in the book.

What kind of problems do you find in this book? Mathematical problems for which no straightforward method of solution offers itself. Problems where you need an idea. Problems that lead to an interesting mathematical concept. Some chapters introduce an area of mathematics, for example graph theory in Chapter 4 or number theory in Chapter 8. These topics are also introduced via a sequence of problems, and at places where it is natural to state a theorem and its proof, the proof is developed from the point of view of a problem-solver.

Here is a more detailed outline. In Chapters 1 and 2 you will find problems where you count things. Counting is the most basic mathematical activity. By learning to advance step by step you will soon have the wonderful experience of discovering mathematics yourself. An important technique for counting is to use recurrence relations, which are discussed in Chapter 2. The basic idea, called recursion, is to reduce a problem to a smaller problem of the same kind. This is also behind the principle of mathematical induction introduced in Chapter 3. Using induction you will prove EULER's famous formula in Chapter 4 and then use it to show a remarkable impossibility result: you cannot connect five points in the plane pairwise so that the connecting curves don't intersect. That we can prove such an assertion rigorously is actually quite amazing: how can we control the infinitely many ways to draw the points and curves? After this excursion into graph theory we return to counting in Chapter 5, this time formulating general systematic counting principles which are often useful.

Chapters 6 and 7 are central: Chapter 6 introduces general problem-solving strategies, that is, strategies which are useful beyond mathematics. Chapter 7 is about logical foundations and the most important general types of proof. These are illustrated by many examples. Chapter 8 is about number theory, that is, prime numbers, divisibility etc.

Here you can use the various types of proof to answer interesting questions about numbers. Number theory also provides useful tools and pretty examples for the following chapters. In Chapter 9 you learn about the pigeonhole principle, a simple idea which spawns amazing consequences when used aptly. Chapters 10 and 11 introduce the extremal principle and the invariance principle. These are versatile tools for solving mathematical problems, and also turn up again and again in all exact sciences. Here you also learn about the fundamental notions of permutations and their signature.

Sections titled *Going further* at the end of Chapters 4, 5, 6, 10 and 11 encourage you to delve deeper into the matters touched upon in these chapters and give you a glimpse of the higher summits of mathematics. Occasionally you may find here an unfamiliar term without detailed explanation, but you should read on, just to get an idea of what else mathematics has to offer. The references at the end of the book guide you to literature that will be useful for further exploration.

The mathematical topics treated in the book are elementary, that is, they can be understood with very few prerequisites. However, we take a higher perspective than you may be used to: often problems are formulated in greater generality than is common at school, we pay attention to logically complete arguments, and although the tone is generally informal, we use modern mathematical terminology, in particular the language of sets and maps. This is explained in Appendix B.

The chapters are mostly independent of each other, so you may read them in a different order. However, later chapters tend to be more demanding than earlier chapters.

Hints for using the book

You learn problem-solving by solving problems, and by understanding other people's solutions. You learn proof by proving, and by carefully reading and understanding proofs. The problems in the text and the exercises at the end of each chapter give you ample opportunity for this. Be an active reader! After reading a problem, first try to solve it yourself. After reading part of the solution, put the book aside and think about how to go on. The symbol

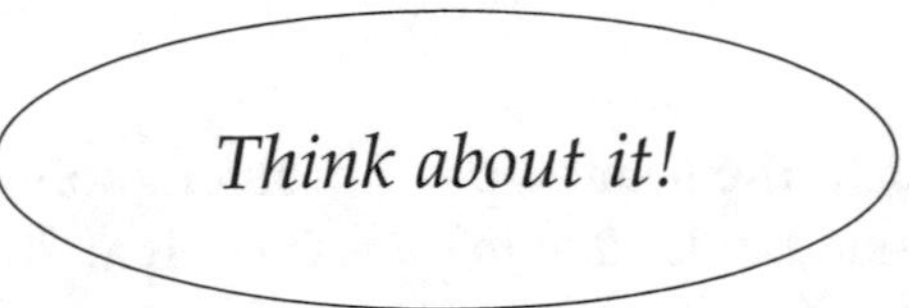

is meant to remind you of this. Whenever you read a statement, ask yourself whether you are convinced that it is true. Keep paper and pencil handy. Ask yourself: why do it like that, couldn't I do it better?

Explain your solution, or the solution you have just read, to a friend. Let him or her play the devil's advocate, who doesn't let you get away with the slightest imprecision, who tries to find gaps in your argument, who asks after each step: "But what if ...". After a while you will learn to be your own devil's advocate.

Sometimes you will have other ideas for solutions than what is given in the text. Every problem has many solutions, and also the same idea can be formulated in many ways. Try to work out whether your solution has the same core idea or whether it is completely different.

The exercises at the end of each chapter are a particularly important part of the book. Work on them. It will pay off. You will find hints for some of the exercises at the end of the book. But you will profit more from an exercise if you don't immediately succumb to the temptation to look them up.

Each exercise has been assigned a level of difficulty. This should be understood as a rough guideline, as there is no objective way to judge difficulty. Whether and when you find a solution depends on many factors. The levels are as follows:

1 – easy, often solvable in your head

2 – should be doable when you have worked through the chapter

3 – requires more commitment and ideas

4 – difficult

Most problems in the book are treated in four steps:

1. Understanding the problem
2. Investigating the problem
3. Writing up the solution properly

4. Review

These are similar to the four steps of problem-solving introduced by Pólya in his classic book *How to solve it* (Pólya, 2014). Steps 1. and 2. are closely interwoven and carried out in the sections marked by the symbol . The solution is marked by the symbol !, the review by the symbol . The end of each of these sections is marked by the same symbol, and the end of a proof is marked q. e. d. (quod erat demonstrandum = which was to be proven).

Notes for the teacher

This book arose from a course whose main aims were to let students see mathematics from a new perspective (not rules to be followed but a world to be discovered), to give them confidence that they can discover mathematics for themselves and to lead them to appreciate proofs.[2] Along the way students acquire tools for problem-solving, get to know the main types of proofs, learn to write solutions and proofs properly, learn about fundamental ideas that occur everywhere in mathematics, and also pick up important concepts like graphs, congruences, permutations and their signatures. These are introduced in informal, often playful contexts, so they can be shared with friends and will be remembered more vividly than if learned in more abstract ways.

If you want to teach such a course my main piece of advice is: *Less is more.* Don't try to 'cover' a certain curriculum. Take time to solve problems together with the students in class, to develop ideas for solutions, to try out different approaches, to suffer the frustration of getting stuck and enjoy the satisfaction of success together. Share with the students your experience of how to approach a mathematical problem, rather than presenting prefabricated solutions. *Always start with an easy problem.* There will be enough challenges for everyone.

I am happy to acknowledge the help and valuable feedback that

[2]This course, titled *Mathematisches Problemlösen und Beweisen* (Mathematical problem-solving and proving), was introduced in the mathematics curriculum at Carl von Ossietzky University of Oldenburg, Germany, in 2011, with the goal of smoothing the difficult transition from high school mathematics to university mathematics.

I got from many students, colleagues and readers of the first German edition. I want to thank especially the student tutors Stefanie Arend, Simone Barz, Karen Johannmeyer, Marlies Händchen, Stefanie Kuhlemann, Roman Rathje, Kathrin Schlarmann, Steffen Smoor and Eric Stachitz, and my colleague Andreas Defant. Sunke Schlüters contributed many excellent ideas for additional problems and helped me with many of the pictures in the book. Sophy Darwin made many valuable suggestions beyond her thorough language editing. My deepest gratitude is due to my wife Ricarda Tomczak for continuous encouragement and innumerable discussions and suggestions.

Now I wish you many joyful hours of solving problems and thinking about proofs, of discovering the beauty of mathematics. If you have any comments or suggestions please write me an email at daniel.grieser@uni-oldenburg.de.

Oldenburg,
February 2018 *Daniel Grieser*

1 First explorations

We begin our journey into mathematics by investigating three problems. The first one is a simple warm-up exercise, but the other two require some serious searching before we find a solution. During this search we will observe ourselves: How do we proceed intuitively when solving a problem? At the end of the chapter we will collect our strategies into a toolbox. This will be of great value when we attack harder problems.

1.1 Cutting up a log

? Problem 1.1

You have a log which is 7 metres long. How long does it take to cut the log into pieces one metre long, assuming that a single cut takes half a minute?

This is easy! Your solution probably looks something like this.

! Solution

You need six cuts, therefore it takes six times half a minute, that is three minutes. !

Review

Let us have a look at our solution. We observe some basic strategies.

- We first found the number of pieces (seven) and then the number of cuts (six). This is a simple example of an **interim goal**.
- A typical beginner's mistake is to calculate like this: Divide 7 metres by 1 metres, therefore we need 7 cuts. But the correct answer is that 7 pieces require 6 cuts. This is a typical **shift by one**. What do you do when you are not sure whether you should shift by one or not? A **sketch** will help, see Figure 1.1.

D. Grieser, *Exploring Mathematics*, Springer Undergraduate Mathematics Series, https://doi.org/10.1007/978-3-319-90321-7_1

You should keep an eye out for shifts by one. See exercise E 1.1 for more examples.

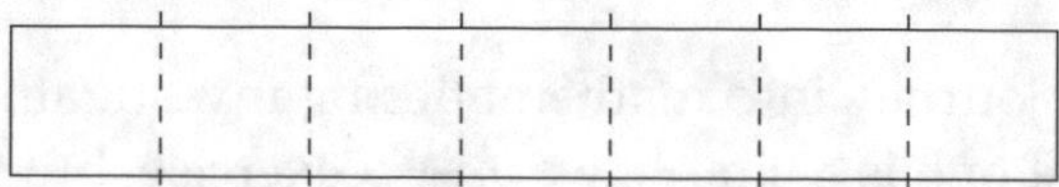

Figure 1.1 The number of cuts is one less than the number of parts

- To make this still clearer, we could **simplify** the problem: we replace 7 by 2 or 3 and observe what happens: one cut for 2 metres, 2 cuts for 3 metres etc.
- We recognize a **pattern:** *The number of cuts is always one less than the number of parts.*
- How can we be certain that this is *always* the case[1]? We need a *proof.* This could run as follows.

  Every cut has a part immediately to its left. All parts arise in this way except the last one at the right-hand end. Therefore the number of parts is equal to the number of cuts plus one.
- We could save some cuts by cutting first into parts 3 and 4 metres long, putting these two pieces next to each other and then cutting both simultaneously, and then continuing in a similar manner. Although loggers certainly wouldn't do this, it is still an interesting problem what the smallest number of cuts is in this case. See exercise E 1.5.

1.2 A problem with zeroes

Let us try a harder problem.

? Problem 1.2

How many zeroes are at the end of $1 \cdot 2 \cdot 3 \cdot \cdots \cdot 99 \cdot 100$?

[1]Also for a million pieces, for example; we cannot draw a picture for this. Although the problem was not worded in this generality, it is satisfying to recognize general patterns.

One writes 100! (in words: one hundred **factorial**) for the product $1 \cdot 2 \cdot 3 \cdot \dots \cdot 99 \cdot 100$. Zeroes at the end of a number we call **trailing zeroes.**

Before reading on, think about the problem yourself.

Investigation

We cannot calculate the product, it is too big even for a calculator.[2] We need an idea how to do it a different way. For a moment you might think the answer is two: the two zeroes of the factor 100. Looking more closely, you might discover that the product also contains the factor 10, and also 20, 30 and so on, each contributing another trailing zero. So we get 11 zeroes. Have we found them all? Can we be sure that 11 is the solution?

To answer this we need to understand where the trailing zeroes come from. We could proceed as follows.

▷ **Simplify!** This means that we consider the same problem for smaller numbers than 100, in order to **get a feel for the problem.** For example, $2! = 1 \cdot 2 = 2$, $3! = 1 \cdot 2 \cdot 3 = 6$, $4! = 1 \cdot 2 \cdot 3 \cdot 4 = 24$, $5! = 1 \cdot 2 \cdot 3 \cdot 4 \cdot 5 = 120$.

▷ For easier bookkeeping we make a **table:**

n	1	2	3	4	5	6
$n!$	1	2	6	24	120	720

The first trailing zero arises with 5!. Why? Write $5! = 1 \cdot 2 \cdot 3 \cdot 4 \cdot 5$. **Where does the trailing zero of the result 120 come from**? Think about it!

A trailing zero means that 120 is divisible by 10. Now $10 = 2 \cdot 5$, and both 2 and 5 are factors in 5!. That's where the zero comes from. In $4! = 1 \cdot 2 \cdot 3 \cdot 4$ there is no factor 5, therefore it has no trailing zero.

Important insight: A number has a trailing zero if it is divisible by 2 and by 5.

[2]Some modern calculators can do this. If yours does, try to make it solve the same problem for 1000! or 10000!. Or Exercise E 1.6.

▷ How about two trailing zeroes? Before reading on, try to predict which is the smallest n whose factorial has two trailing zeroes. Use the insight above.

To have two trailing zeroes, a number must be divisible by 100, which is $10 \cdot 10 = 2 \cdot 5 \cdot 2 \cdot 5 = 2^2 \cdot 5^2$. So we need two factors 5 and two factors 2. Let us imagine continuing the table, without actually doing any calculation. **We only need to watch for fives and twos.** Where is the next factor 5? Not in 6, 7, 8, 9, but in 10 since $10 = 2 \cdot 5$. There is a surplus of factors 2; each one of 2,4,6,... contributes at least one. Just two of them would be enough. Therefore 10! has two trailing zeroes. How does this continue? Try to solve the problem now.[3]

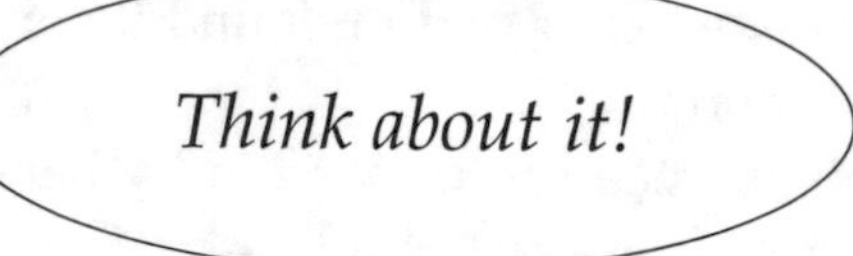

▷ Maybe you recognize the **pattern:** Let k be any natural number. For a number to have at least k trailing zeroes it must be divisible by $10^k = 2^k \cdot 5^k$, that is, it must contain k factors 5 and k factors 2.

▷ So we should count how often the factors 5 and 2 appear in 100!.

Let us first look at the factor 5: Which factors $1, 2, 3, \ldots, 100$ in 100! contribute 5? Clearly $5, 10, 15, 20, 25, \ldots, 100$ do. That's how many? If we write these numbers as $1 \cdot 5$, $2 \cdot 5$, $3 \cdot 5, \ldots$, $20 \cdot 5$ then we see that there are 20 of them. Is this the solution?

Watch out! $25 = 5 \cdot 5$ contributes two factors 5. Which other factors in $1 \cdot 2 \cdot 3 \cdot \cdots \cdot 100$ do that? Clearly the multiples of 25, that is, 25, 50, 75, 100. Each of them contributes another 5, that's 4 additional fives. (The factor $125 = 5^3$ would contribute two additional fives, but we don't get this far in 100!.) All in all the factor 5 appears 24 times in 100!.

How about the factor 2? Clearly, each one of the 50 even numbers $2, 4, \ldots, 100$ contributes at least one factor 2, so we get more than 24 such factors. But the surplus twos (those beyond the 24th) do not yield a trailing zero in 100! since they have no partner 5.

[3] Whenever you see the following symbol, put the book aside, take pen and pencil and try to work out the problem, or at least the next step, by yourself.

We obtain the *solution:* 100! has 24 trailing zeroes. 

To tell others about our solution we should write it up well. This will also help us to check it for completeness. This could look as follows.

! Solution of Problem 1.2

The number of trailing zeroes of 100! is the biggest integer k for which 100! is divisible by 10^k. Because $10 = 2 \cdot 5$, this is the biggest integer k for which 100! is divisible by $2^k \cdot 5^k$, that is, by 2^k and by 5^k.

The factor 5 arises as a factor in the 20 numbers 5,10,15,...,100, and twice in the 4 numbers 25,50,75,100. Therefore it arises in 100! precisely 20+4=24 times, so $k = 24$ is the largest k for which 100! is divisible by 5^k.

The factor 2 arises as a factor in the 50 numbers 2,4,...,100, in some of them more than once. Therefore it arises in 100! more than 50 times. The precise number does not matter: we only need that the number is at least 24.

Therefore the largest k for which 100! is divisible by 2^k *and* by 5^k is $k = 24$. Therefore 100! has precisely 24 trailing zeroes. !

Review of Problem 1.2

We were able to solve the problem by **understanding the mechanism** that produces trailing zeroes. We made use of the following problem-solving strategies.

First we **got a feel for the problem** by the following steps.

- ❑ **Simplify** the problem to make it tractable (consider 5! instead of 100!).
- ❑ Make a **table** for easier bookkeeping.

The breakthrough came with more considerations:

- ❑ **Analyse the goal** (where do trailing zeroes come from?), first in the simpler case 5!.
- ❑ From the simpler problem work towards the original problem **step by step** (when does the next trailing zero appear – at 10!); here we focus on what matters (factors 5 and 2).

- ❏ Recognize a **pattern or rule** (number of trailing zeroes = number of factors 5 and 2).

Of course we always need to be alert not to overlook anything (additional fives in 25, 50, 75 and 100). ○

You might like to look at the general question now: How many trailing zeroes does $n!$ have for an arbitrary natural number n? See exercise E 1.6.

1.3 A problem about lines in the plane

? Problem 1.3

Suppose you draw n straight lines in the plane, no two of which are parallel and no three of which meet at a point[4]*. These lines subdivide the plane into regions. How many regions are there?*

In this problem, we write n for an arbitrary natural number: $n = 1, 2, 3, \ldots$. You might think that it is harder to consider the general problem (any n) than the special problem (one fixed value of n, for example $n = 20$).

But we will see that the more general problem is easier! Why? Recall that in Problem 1.2 we made progress by first considering the same problem with 100 replaced by 1,2,3,4,... Then we were able to solve the *special* problem by understanding the *general* mechanism which produces trailing zeroes. Here things will be similar.

For this reason we will pose many problems in a general form from now on.

Investigation

▷ Let us have a close look at the problem: Do we understand the assumptions? Which configurations of lines are permitted, which

[4] If both of these conditions are satisfied then we say that the lines are **in general position**.

are not? We make a few **sketches**, look at **examples**. The configurations in Figure 1.2 are permitted, those in Figure 1.3 are not: in the left-hand picture two lines are parallel, in the right-hand picture three lines meet at a point.

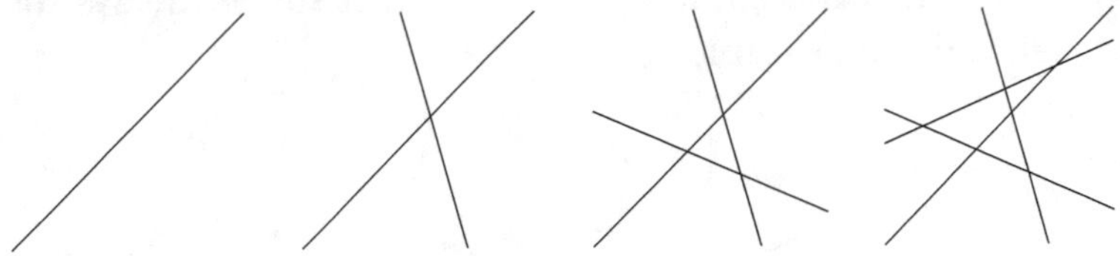

Figure 1.2 Permitted configurations

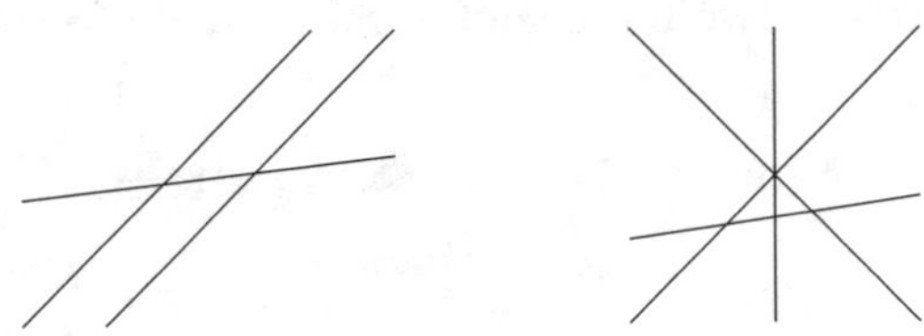

Figure 1.3 Configurations which are not permitted

▷ What are we looking for? The number of regions created by the lines. Let us count the regions in each of the configurations in Figure 1.2, where $n = 1, 2, 3, 4$, and make a **table.**

n	1	2	3	4
a_n	2	4	7	11

Here we introduced the notation a_n for the number of regions, so $a_1 = 2$, $a_2 = 4$ and so on (**introduce notation**).
To get more familiar with the problem we also look at the configurations in 1.3, although they are not permitted: In the left-hand configuration we have $n = 3$ with 6 regions, in the right-hand configuration we have $n = 4$ with 10 regions. These numbers differ from the values of a_3, a_4 in the table: Apparently, the assumptions in the problem actually matter!

▷ *Attention:* It may well be that, even for permitted configurations, the number of regions depends on the *position* of the lines, not just on the *number* of lines. To check this out we draw a few other configurations with $n = 4$, see Figure 1.4. We see that we get 11 regions in each case, as before. This does not prove anything, but it is reassuring: Maybe the number of regions is independent of the position of lines after all (for permitted configurations, of course)! It is part of the problem to find out whether this is always the case, not just in these examples.

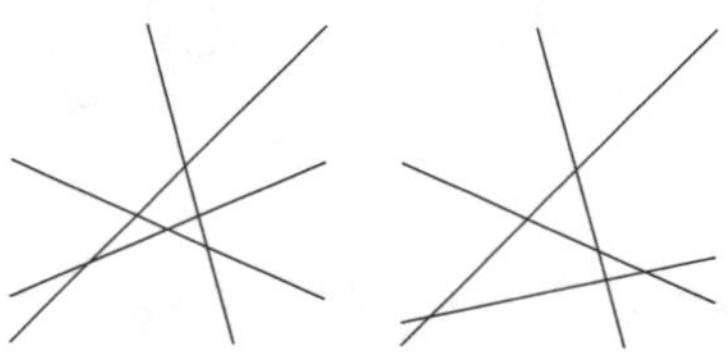

Figure 1.4 Two more configurations with four lines

▷ Look at the table for a_n. Can you see a **pattern?**

Maybe you discovered one of the following patterns:

- Every number a_n is the sum of the numbers directly to its left and above it.
- As we go from left to right, the difference of consecutive numbers a_n increases by one.
- Every a_n is the sum of the numbers in the first row, up to n, plus one.

All are correct. **There are always several ways to proceed.**[5] Maybe you have discovered yet another pattern.

To proceed let us suppose you had found the first pattern. How can we write it as a general formula? The number to the left of a_n is a_{n-1}, the number above it is n. So we observed that

$$a_n = a_{n-1} + n \quad \text{for } n = 2, 3, 4. \tag{1.1}$$

[5] Actually these descriptions are closely related. Convince yourself that the first two say essentially the same thing, and that the third one can also be related to them.

Does this rule continue to hold for higher n? Of course we could check a few more examples. But can we be sure that it holds for *all* n? We need to find a general argument. This is an **interim goal** in our quest to solve Problem 1.3.

▷ Let us try to find a reason why equation (1.1) should hold for all n. What does this equation mean in terms of our original problem? (We **analyse our goal.**)

Recall that a_n is the number of regions for n lines, so a_{n-1} is the number of regions for $n-1$ lines. So (1.1) says: Whenever we add a line to already existing $n-1$ lines, the number of regions increases by n. Why should this be true? Think about it!

If you don't see it right away, it is useful to consider an **example,** always looking out for a **general rule.** Suppose we start with two lines and then add a third one, which we call ℓ. How do new regions arise? For simplicity let us colour the previously existing lines black and the new line red. Compare the pictures. Take your time.

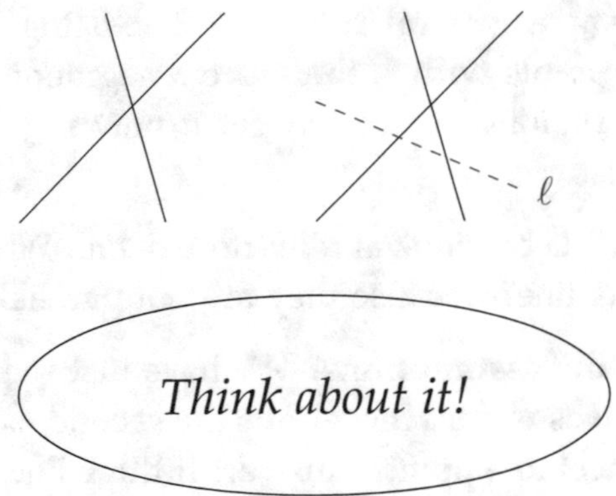

Think about it!

▷ *Insight:* The new regions arise because the new line ℓ cuts through some of the previously existing regions.

▷ How many previously existing regions are cut by ℓ? Well, three, but can we relate this to other aspects of the figure (remember to look for general rules; 'three' is not a general rule)? Is there another 'three' in the picture? Yes: ℓ is cut into three segments by the black lines. And clearly this is a **general rule:** The number of regions cut by ℓ is the same as the number of segments of ℓ.

What's next? Remember what we are after: Adding a new line to

$n-1$ existing (black) lines adds n regions. So we need to relate the *number of new regions* to the *number of black lines*. We saw: The number of new regions equals the number of segments of ℓ. Can you relate the *segments* of ℓ to the *black lines*? Think about it.
Of course! Each black line intersects ℓ, and the number of segments of ℓ must be one more than the number of intersection points! In the example: two black lines, so two intersection points, so three segments.[6]

▷ To get comfortable, let us check this with three black lines, adding a fourth one: Three intersection points, four segments, therefore four new regions.

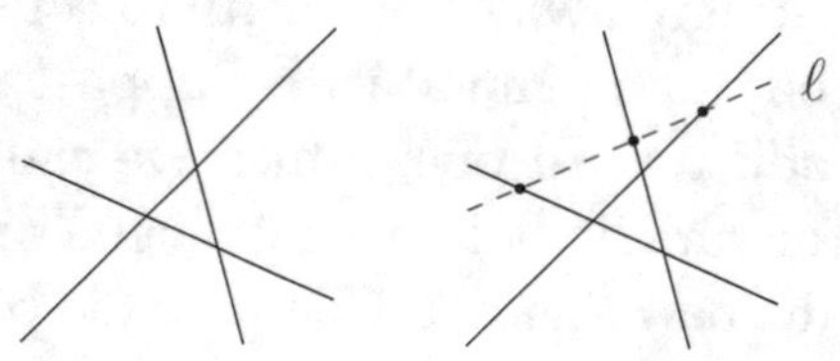

▷ Clearly, this is a general rule: $n-1$ existing lines yield $n-1$ intersection points with ℓ, therefore n segments of ℓ, therefore ℓ cuts n 'old' regions, hence we get n new regions. This proves equation (1.1)!

▷ Does it really? Let us look at it in more detail. Why does ℓ intersect every existing line? Because they are not parallel!

▷ **Did we use the assumptions?** We have just seen where the first assumption was used. How about the second assumption, i.e. no three lines meet in a point? Answer: If three lines were allowed to meet in a point then ℓ might pass through the intersection point of two black lines, so there would be less than $n-1$ intersection points, hence less than n new regions.

▷ So we have reached our interim goal: We proved equation (1.1). This type of equation, which relates a_n to a_{n-1}, is called a **recurrence relation**. A recurrence relation is quite useful, it allows us

[6]This is just like the cuts and pieces of the log in Problem 1.1.

to calculate $a_2, a_3, a_4 \ldots$ etc. quickly (starting with $a_1 = 2$), without drawing pictures.[7]

▷ But really we would like a *closed formula*, i.e. a formula which allows us to calculate a_n directly from n, without first calculating $a_1, a_2, \ldots$. We obtain this in two steps: In the first step, we *plug the recurrence relation into itself repeatedly:* First note that replacing n by $n-1$ in (1.1)[8] yields $a_{n-1} = a_{n-2} + (n-1)$, so

$$a_n = a_{n-1} + n = a_{n-2} + (n-1) + n.$$

Continuing in the same manner we get

$$\begin{aligned} a_n &= a_{n-2} + (n-1) + n = a_{n-3} + (n-2) + (n-1) + n = \ldots \\ &= a_1 + 2 + 3 + \cdots + n \\ &= 2 + 2 + 3 + 4 + \cdots + n\,. \end{aligned}$$

(How do we know what to put in the second line? The previous two expressions exhibit a pattern: The first number added is one more than the index of the a before it, for example $a_{n-2} + (n-1) + \ldots$, $a_{n-3} + (n-2) + \ldots$. So when we are down to a_1 then we start $a_1 + 2 + \ldots$.)

▷ This last expression is still not satisfying because of the dots: For example, if $n = 100$ then we would have to add one hundred numbers to get a_{100}. So in a second step we use the following famous trick[9] [10] for calculating $1 + \cdots + n =: s$ which then yields

---

[7] And, incidentally, it shows that the number of regions depends only on the number of lines, not on their (permitted) position.

[8] We are allowed to do this since (1.1) is true for all numbers n, so in particular for the number $n-1$.

[9] You should put that in your bag of tricks! It is usually attributed to Carl Friedrich Gauss (1777-1855), but most likely it is much older.
By the way, the original spelling of this name is Gauß. The German letter ß is a sharp s.

[10] Some people prefer to proceed as follows: In $1 + \cdots + n$ first add the first and last summand to obtain $n+1$, then add the second and second to last summand to get $2 + (n-1) = n+1$ again and so on. This works fine except that you have to think about what happens when you get to the middle, and this is different for even and odd n. Doing each of these cases separately you get the formula in the text also. But the method in the text is better since you don't need to distinguish two cases.

a quick formula for a_n. Write s two times and add:

$$\begin{array}{ccccccccccc} 1 & + & 2 & + & \dots & + & (n-1) & + & n & = & s \\ n & + & (n-1) & + & \dots & + & 2 & + & 1 & = & s \\ \hline (n+1) & + & (n+1) & + & \dots & + & (n+1) & + & (n+1) & = & 2s \end{array}$$

How many times does $n+1$ appear in the last line? The first line shows that the answer is n. Hence $n(n+1) = 2s$, so

$$1+2+\cdots+n = \frac{n(n+1)}{2}, \tag{1.2}$$

and to get a_n we simply add one. This is the closed formula we were looking for.

Let us write up the solution properly.

! Solution of Problem 1.3

Obviously $a_1 = 2$. We first prove that for all $n \geq 2$ the recurrence relation

$$a_n = a_{n-1} + n \tag{1.3}$$

holds. To prove this, suppose $n-1$ lines are given in the plane in general position. Add another line ℓ which is not parallel to any of the given lines and does not pass through any of their intersection points. Then ℓ intersects every one of the $n-1$ given lines, and we get $n-1$ intersection points on ℓ. These points subdivide ℓ into n segments. Each segment splits one of the regions formed by the $n-1$ lines into two, so adding ℓ produces n additional regions. This proves equation (1.3).

Plugging the recursion into itself repeatedly and using equation (1.2) we get the *solution*

$$a_n = 2+2+3+4+\cdots+n = 1+\frac{n(n+1)}{2}. \tag{1.4}$$

!

Review of Problem 1.3

As in Problem 1.2 we first got a feel for the problem with the help of **examples** (in this case **sketches**) and a **table**. The examples also helped us to **understand the assumptions in the problem.**

Then we found a **pattern. Introducing some notation** helped us to express the pattern as an equation (1.1). To investigate whether the pattern continues to hold for all n and all configurations, we interpreted the equation in terms of the original problem (adding a line). We **focussed on the goal** (where do the new regions come from?). We could finally prove the formula by **advancing in small steps**, always keeping the goal in mind. Here again it was useful to look at an example, but we were careful to argue in a way that would work in general. A **look at the assumptions** helped us to fill in small gaps in the argument.

Finally we used the GAUSS trick (a **technique**) to derive the solution (1.4) from the recurrence relation (1.1).

1.4 Toolbox

In this chapter you encountered the most important general problem-solving strategies. We now collect them into a toolbox, which will be supplemented in later chapters. You can find the complete tool box in Appendix A.

1. Understand the problem

Read the problem carefully. What is given, what are we looking for? What are the assumptions?

2. Investigate the problem

- ❑ Get a feel for the problem, get your hands on it.

  Useful strategies for this:

  - First consider a simpler problem
  - Look at special cases, examples
  - Make diagrams and tables

- ❑ Use the examples to find a general rule
- ❑ Look for patterns
- ❑ Set yourself interim goals
- ❑ What is essential?

❑ Work step by step

❑ Check if you used the assumptions

❑ Introduce notation[11]

3. Write up your solution properly
Pay particular attention to:

❑ Complete arguments

❑ Sensible structure

❑ Understandable writing (mention ideas, motivations)[12]

❑ Correct use of mathematical language

4. Review
What have we learned? Is the solution reasonable? Could it be done differently, better?

Writing up the solution properly and reviewing helps you to check your solution. Sometimes you notice gaps in your solution during these steps. The review also helps for other problems.

Exercises

When working on the exercises, think about which problem-solving strategies might be useful.

1 E 1.1 Here are some examples of shifts by one. Or aren't they? You may use fingers to count, but also try to find a good argument.

- You see a row of five trees, each one 20 meters from the next. How long is the row? What if there are 20 trees?
- I booked a hotel room from May 20 (arrival) to May 23 (departure). That's how many nights? How many from May 3 to May 29?

[11]That is, short identifiers, for example a_n for the number of regions in Problem 1.3.

[12]Many people think that mathematical texts consist only of equations and symbols. This is far from true. Explanations are as important as equations, and in some mathematical texts there are no equations at all.

- You see a row of 5 trees. The row is 100 meters long. What's the distance between successive trees, supposing it's always the same?
- I start work at 8 am, I work until 5 pm. That's how many hours?
- The clock on my office wall chimes every hour, on the hour. How often do I hear it chime during a working day?
- How many three digit natural numbers are there?
- How many integers n satisfy the inequalities $15 \leq n \leq 87$? How many satisfy $-10 \leq n \leq 10$?

E 1.2 You want to arrange matches into n squares as follows. How 1
many matches do you need? The side length of the squares is one match.

E 1.3 Let $n \in \mathbb{N}$. Consider the sum of the first n odd natural 2
numbers, that is,

$$1 + 3 + 5 + \cdots + (2n - 1).$$

Calculate the sum for some values of n. Can you see a pattern? Make a conjecture and prove it.

E 1.4 Let $n \in \mathbb{N}$. Is it possible to draw n points $P_1, \ldots, P_n$ and a 2
straight line g on a sheet of paper such that none of the points lies on g, but g intersects each of the line segments $P_1P_2, P_2P_3, \ldots, P_{n-1}P_n, P_nP_1$? Look at some examples, make a conjecture and prove it.

E 1.5 Modify Problem 1.1 as follows: suppose that before any cut 3
you are allowed to put an arbitrary number of pieces obtained in earlier cuts next to each other, and to make one straight cut through all of them. Now how many cuts do you need? How many cuts for a log of length 4, 8, 16, 32, 27, n? Give a reason why your answer is best possible, i.e. why it cannot be done with fewer cuts.

E 1.6 Let $n \in \mathbb{N}$. Find a formula for the number of trailing zeroes 2
of $n!$.

E 1.7 The strategy 'special cases, looking for patterns' can be useful 2
for finding formulas. Here are some examples.

a) Find a closed formula for the sum

$$1 \cdot 1! + 2 \cdot 2! + \cdots + n \cdot n!$$

b) Find a closed formula for the sum

$$\frac{1}{2!} + \frac{2}{3!} + \frac{3}{4!} + \cdots + \frac{n}{(n+1)!}$$

c) (With more calculation:) Compute the numbers

$$1 + \frac{1}{1^2} + \frac{1}{2^2},\ 1 + \frac{1}{2^2} + \frac{1}{3^2},\ 1 + \frac{1}{3^2} + \frac{1}{4^2}, \ldots$$

as fractions. What do you observe? Formulate and prove a general formula. Use this formula to find a closed formula for the sum

$$\sqrt{1 + \frac{1}{1^2} + \frac{1}{2^2}} + \sqrt{1 + \frac{1}{2^2} + \frac{1}{3^2}} + \cdots + \sqrt{1 + \frac{1}{n^2} + \frac{1}{(n+1)^2}}$$

1-2 E 1.8 For each of these sequences find a pattern and formulate it in words, then as a formula, where the numbers are $a_1, a_2, \ldots$. Invent more such problems.

Example: 1, 4, 9, 16, 25: square numbers, $a_n = n^2$

Example: 7, 9, 12, 16, 21: the difference increases by one at each step, starting at 2 from a_1 to a_2; formula: $a_n = a_{n-1} + n$ with $a_1 = 7$. Another (closed) formula is $a_n = \frac{n(n+1)}{2} + 6$.

a) 3, 4, 5, 6, 7

b) 3, 9, 36, 180, 1080

c) 1, −1, 1, −1, 1

d) 1, 1, 1, 3, 5, 9, 17, 31

e) 2, 8, 24, 64, 160

3 E 1.9 For $n \in \mathbb{N}$ denote by s_n the number of ways to write n as ordered sum of natural numbers. Count n itself (one summand) as one of these ways. For example

$$2 = 1 + 1 \text{ and } 3 = 1 + 2 = 2 + 1 = 1 + 1 + 1,$$

so $s_2 = 2$ and $s_3 = 4$. Find s_4 and s_5, conjecture a general rule and prove it.

E 1.10 Make a conjecture on the truth of the statement '$n^2 + n + 41$ is 2
a prime number for all $n \in \mathbb{N}$.' Then check your conjecture by giving a proof or a counterexample for the statement.

E 1.11 Investigate: Suppose you draw n straight lines in the plane, no 4
three of which meet in a point. How many regions do you get? Note that lines are allowed to be parallel. The answer will not only depend on n. What else will it depend on? Identify quantities associated with a configuration of lines that allow you to calculate the number of regions, and find a formula for this number in terms of these quantities. In the special case of lines in general position you should get the formula found in this chapter.

2 Recursion – a fundamental idea

Have you ever seen a Russian Matryoshka doll? When you open up this wooden figure then you will find a smaller figure inside which looks just like the first one. You can open this figure again and find a yet smaller one, and so on.

Some mathematical problems can be tackled in a similar way: solve the problem by reducing it to a smaller problem of the same kind. This technique is called recursion. It is frequently used for counting problems, and the equation which expresses this reduction is called a recurrence relation. You have encountered this already in Problem 1.3. We will now explore the method systematically and see some more examples.

For any given problem there are typically two tasks: First, find a recurrence relation. Second, solve the recurrence relation; that is, find a closed, non-recursive formula. In this chapter you will learn about ways to tackle both steps.

2.1 Recursion in counting problems

Often the idea of recursion is expressed in terms of a recurrence relation:

Definition A **recurrence relation** for a sequence of numbers $a_1, a_2, a_3, \ldots$ is an equation which expresses any a_n in terms of $a_{n-1}, a_{n-2}, \ldots, a_1$, and n.

When solving Problem 1.3 we encountered the recurrence relation $a_n = a_{n-1} + n$. This expresses a_n in terms of a_{n-1} and n only, the previous values $a_{n-2}, \ldots$ are not used. We call this a *recurrence relation*

D. Grieser, *Exploring Mathematics*, Springer Undergraduate Mathematics Series, https://doi.org/10.1007/978-3-319-90321-7_2

of length one. Here are some more examples of recurrence relations:

$$a_n = a_{n-1} + a_{n-2}$$
$$a_n = n a_{n-1}$$
$$a_n = 1 + a_1 + \cdots + a_{n-1}$$

When writing such equations we always mean that they should hold for all n where all occurring indices are at least one. So the first equation means

$$a_3 = a_2 + a_1, \; a_4 = a_3 + a_2, \; a_5 = a_4 + a_3 \text{ etc.}$$

Recurrence relations are a useful technique for solving **counting problems**. In a counting problem we are interested in certain configurations or objects and want to know how many there are.

Examples of counting problems

- How many regions are formed by n straight lines in the plane that lie in general position (see Problem 1.3)?
- In how many ways can you choose 6 numbers out of $1, \ldots, 49$?
- Suppose that n people meet, and everyone shakes hands with everyone else – but only once. That's how many handshakes?
- Suppose you have a board of size 2×10, and 10 domino pieces. How many ways are there to cover the board with the dominoes (see Problem 2.2)?

Often the problem involves a number n which indicates the 'size' of the problem. Let us denote the number we are looking for by a_n. Then we are looking for a formula that allows us to compute a_n directly from n. The recursion technique runs as follows.

Technique: Recursion

1. **Look for a recurrence relation:** Can you reduce the problem of size n to a smaller problem *of the same kind*? If yes then you

will obtain a recurrence relation for the sequence of unknown numbers a_n.

2. **Solve the recurrence relation:** Use the recurrence relation to find a formula for a_n. To do this you also need an *initial condition*, that is, the values of a_n for some small n (depending on the recurrence relation this could be one, two or more values).

This is precisely how we proceeded in Problem 1.3: Removing one of the n given lines, we reduced the problem to a smaller problem of the same kind (with $n-1$ lines). Arguing that adding the line back in will produce n extra regions, we derived the recurrence relation $a_n = a_{n-1} + n$. Then we solved the recurrence relation using the Gauss trick.

Looking for a recurrence relation is an example of setting an **interim goal:** Instead of aiming for a formula for a_n directly, we content ourselves first with the more modest goal of finding a recurrence relation.

The two steps of the recursion technique are very different in character, as the examples in the next sections will show. There is no general technique for solving recurrence relations. Simple recurrence relations can be solved simply by repeatedly plugging them into themselves, as we did in Problem 1.3. For many recurrence relations this does not help, however. An example is the famous Fibonacci recurrence relation, and in Section 2.4 you will learn how to solve this and similar recurrence relations. There are also recurrence relations which cannot be solved at all by a closed formula.

Recursion is only one way to approach counting problems. You will learn about other counting principles in Chapter 5.

Remark

When you are given a problem for some fixed number of objects (for example, 100 lines in the plane), it may be useful to generalize it (n lines in the plane). Only then you will be able to use the recursion technique!

This is a curious phenomenon, typical for mathematics, and for science in general:

Sometimes a problem simplifies when we look at a more general problem.

This sounds like a paradox since you would expect a more general problem to be harder. But this is precisely how we were able to solve Problem 1.2: Instead of looking at 100! as required, we looked for the general way in which zeroes at the end of $n!$ are 'produced'.

This principle is the opposite of the problem-solving technique of looking at special cases. Both are useful!

2.2 The number of subsets

Imagine you have three photographs and want to give some selection of them to a friend. You have not decided yet how many and which photos you want to give her. How many possibilities are there? You could give away one photo (3 possibilities: photo 1 or 2 or 3) or two (3 possibilities: photos 1,2 or 1,3 or 2,3), or all three or none at all (one possibility each). That's $3+3+1+1=8$ possibilities all in all. How many possibilities would you have with 4 or 5 or more photos? Mathematically speaking, we want to find the number of subsets of a set with three or more elements (your photos).

We will use the language of sets. If you are not familiar with this, read Appendix B first.

? Problem 2.1

How many subsets does the set $\{1,2,\ldots,n\}$ *have?*

We always count the empty set $\emptyset$ and the full set $\{1,\ldots,n\}$ among the subsets[1]. For example, the set $\{1,2\}$ has four subsets: $\emptyset, \{1\}, \{2\}, \{1,2\}$.

Investigation

▷ We first need to **get a feel for the problem**. Therefore we look at a few examples and make a **table**. For simplicity we write, for

[1] This may seem artificial at first. But you will see that in this way the solution will be simpler and more elegant. If you don't want to count the empty set then simply subtract one from the result.

example, 12 for the subset $\{1,2\}$.[2] We denote the desired number by a_n.

n	subsets of $\{1,2,\ldots,n\}$	a_n
1	$\emptyset, 1$	2
2	$\emptyset, 1, 2, 12$	4
3	$\emptyset, 1, 2, 3, 12, 13, 23, 123$	8

▷ Maybe you see a pattern in the last column: $a_1 = 2 = 2^1$, $a_2 = 4 = 2^2$, $a_3 = 8 = 2^3$, the powers of 2. So we **conjecture**:

Conjecture: We have $a_n = 2^n$ for all n.

▷ Why should this be true? How can we be sure that this is still true for $n = 100$, say? As a *first attempt* at an understanding let us order the subsets by their size and see how many subsets there are for each fixed size. For $n = 2$ the numbers of subsets with 0,1,2 elements are 1,2,1, respectively; for $n = 3$ the numbers of subsets with 0,1,2,3 elements are 1,3,3,1, respectively. Adding up, we get 4 and 8 subsets; but there is no clear pattern which would help us explain why we *always* (for *all* n) get a power of 2. We need a new idea.

▷ *Second attempt:* Can we solve the problem recursively? Can we reduce the problem of size n to the same problem of size $n-1$? **Can we find a recurrence relation?**

▷ Let us check the **example** $n = 3$: Can we relate the subsets of $\{1,2,3\}$ to the subsets of $\{1,2\}$? Look at the table above.

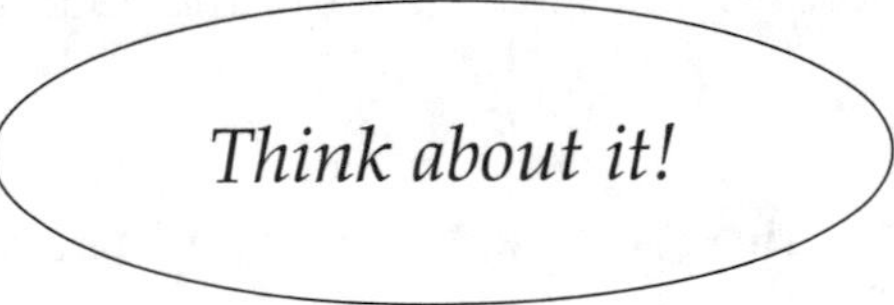

▷ *Observation:* The subset of $\{1,2\}$ are among the subsets of $\{1,2,3\}$: They are precisely those subsets of $\{1,2,3\}$ which don't contain the element 3.

[2]Such **short notation** can be very useful. It should be explained when it is used the first time.

▷ So let us consider the remaining subsets of $\{1,2,3\}$: those that do contain 3 as an element. There are four of them. Can we relate these also to the subsets of $\{1,2\}$? Let us look at them:

$$3, 13, 23, 123 \tag{2.1}$$

and compare them to the subsets of $\{1,2\}$:

$$\emptyset, 1, 2, 12 \tag{2.2}$$

Do you notice anything? Stare at these two lines until you notice something (beyond the fact that each line has four items, of course).

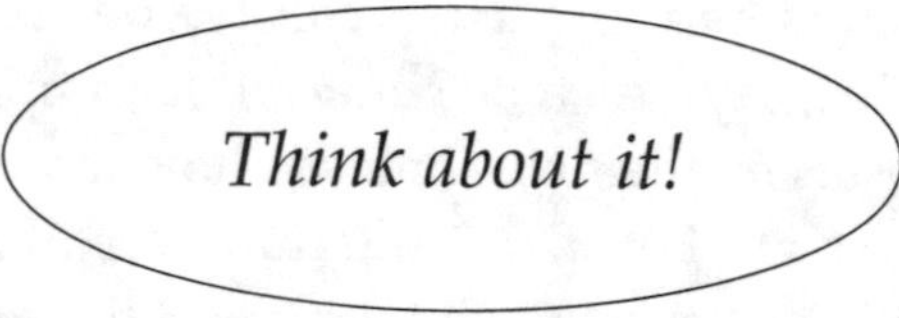

▷ Of course! Each set in row (2.1) arises from a set in row (2.2) by adding the digit 3 to it.

▷ Let us summarize: We have divided the subsets of $\{1,2,3\}$ in two classes: those that contain 3 and those that do not. For each class we showed that it has as many subsets in it as there are subsets of $\{1,2\}$. Therefore $a_3 = a_2 + a_2 = 2a_2$.

▷ The main virtue of this argument is that it **generalizes** to any n: We divide the subsets of $\{1,\ldots,n\}$ in two classes: those that contain n, and those that don't. Each class contains a_{n-1} subsets, by the same reasoning as above. Therefore we get the recursion

$$a_n = 2a_{n-1}.$$

If we plug this into itself repeatedly and use $a_1 = 2$ then we get $a_n = 2^n$ for all n.

We now write this up cleanly, with all the details. This will also help us find out whether the argument has any gaps.

Solution of Problem 2.1

Denote by a_n the number of subsets of the set $\{1,\ldots,n\}$.

1. *Claim:* The recurrence relation $a_n = 2a_{n-1}$ holds for $n = 2,3,\ldots$.

Proof.
We divide the subsets of $\{1,\dots,n\}$ in two classes: the first class contains those subsets that contain n, the second class contains those subsets that do not contain n.

The subsets that don't contain n are precisely the subsets of $\{1,\dots,n-1\}$. So the second class contains precisely a_{n-1} subsets.

For any subset $A \subset \{1,\dots,n\}$ which contains n we consider the subset $B = A \setminus \{n\}$ of $\{1,\dots,n-1\}$. Now any subset $B \subset \{1,\dots,n-1\}$ arises in this way from some A: simply set $A = B \cup \{n\}$. In fact, this is clearly the only choice of A which results in B. Therefore the first class contains as many subsets A as there (*) are subsets B of $\{1,\dots,n-1\}$, i.e. a_{n-1} subsets.

Summing up, we get $a_n = a_{n-1} + a_{n-1} = 2a_{n-1}$ for all $n = 1,2,\dots$.
q. e. d.

2. We now solve the recurrence relation by repeatedly plugging it into itself: If we replace n by $n-1$ in the recursion formula we obtain $a_{n-1} = 2a_{n-2}$. This is allowed for $n \geq 3$ since the recursion holds for *all* numbers $2,3,\dots$, so in particular for the number $n-1$. We plug in: $a_n = 2a_{n-1} = 2 \cdot 2a_{n-2}$. Repeating in a similar vein we get, using the *initial condition* $a_1 = 2$,

$$a_n = 2a_{n-1} = 2 \cdot 2a_{n-2} = \cdots = 2^{n-1}a_1 = 2^{n-1} \cdot 2 = 2^n. \qquad (2.3)$$

The factor 2^{n-1} arises since each time the index of a goes down by 1 we get another factor of 2. The index of a is reduced from n to 1 over all, hence by $n-1$, hence there are $n-1$ factors of 2, so we get 2^{n-1}. !

We now write up the same solution again, but in very formal mathematical language. This language is most useful when writing up complex arguments, so for our problem it is really a bit of overkill. So you may skip this part unless you want to practice formal language. In the second part of the proof we use mathematical induction. This will be introduced in Chapter 3.

! Formal solution of Problem 2.1

Let $T_n = \{A : A \subset \{1,\dots,n\}\}$ be the set of subsets of $\{1,\dots,n\}$ and $a_n = |T_n|$ the number of elements of T_n.

1. *Claim:* We have the recurrence relation $a_n = 2a_{n-1}$ for $n = 2, 3, \ldots$.
Proof.
Let

$$U_n = \{A \in T_n : n \in A\}$$
$$V_n = \{A \in T_n : n \notin A\}.$$

Then

$$T_n = U_n \cup V_n \text{ (disjoint union)}. \tag{2.4}$$

We now prove that U_n and V_n have a_{n-1} elements each. Obviously $V_n = T_{n-1}$, so

$$|V_n| = |T_{n-1}| = a_{n-1}.$$

In order to show that $|U_n| = |T_{n-1}|$ we find a bijection $u : U_n \to T_{n-1}$. Define the map

$$u : U_n \to T_{n-1}, \quad A \mapsto A \setminus \{n\}.$$

In order to show that u is a bijection we write down another map which we then prove to be the inverse of u. Define the map

$$v : T_{n-1} \to U_n, \quad B \mapsto B \cup \{n\}.$$

We claim that v is the inverse of u, that is, that

$$u(v(B)) = B \text{ for all } B \in T_{n-1}, \quad v(u(A)) = A \text{ for all } A \in U_n. \tag{2.5}$$

The first equation holds because of

$$u(v(B)) = u(B \cup \{n\}) = (B \cup \{n\}) \setminus \{n\} = B.$$

Here the first two equality signs are the definitions of v and u, and the third one follows from $n \notin B$. The second equation in (2.5) follows from

$$v(u(A)) = v(A \setminus \{n\}) = (A \setminus \{n\}) \cup \{n\} = A$$

where in the last step we used $n \in A$.
Therefore we have proven (2.5), hence u is bijective. This implies

$$|U_n| = |T_{n-1}| = a_{n-1}.$$

Then (2.4) gives

$$a_n = |T_n| = |U_n| + |V_n| = a_{n-1} + a_{n-1} = 2a_{n-1},$$

which was to be shown. q. e. d.

2. *Claim:* We have $a_n = 2^n$ for all n.

Proof.
We use mathematical induction.

Base case: For $n = 1$ we have $a_1 = 2 = 2^1$.
Inductive hypothesis: Let $n \in \{2,3,\dots\}$ be arbitrary, and assume $a_{n-1} = 2^{n-1}$.
Inductive step: Using the recurrence relation we get

$$a_n = 2a_{n-1} = 2 \cdot 2^{n-1} = 2^n,$$

which was to be shown. q. e. d.

!

Remarks

- Both proofs are mathematically valid. It is a matter of taste which one you prefer. Many mathematics books are written in the formal language, so consider this a useful exercise.
- Note that one needs to argue carefully for the subsets A that contain n: it does not suffice to say that $A \setminus \{n\}$ is a subset of $\{1,\dots,n-1\}$. In addition, you need to show that every subset of $\{1,\dots,n-1\}$ arises precisely once from this operation. In the formal proof this is the statement that u is bijective.
- In order to prove bijectivity of u in the formal proof, one could show that u is both injective and surjective, rather than exhibiting an inverse map. These words are explained in Appendix B. In the informal proof this proof was carried out at the place marked by (*).
  Injective: If $A, A' \in U_n$ satisfy $A \neq A'$ then $u(A) \neq u(A')$.
  Surjective: For every $B \in T_{n-1}$ there is $A \in U_n$ satisfying $u(A) = B$.
- If we interpret $\{1,\dots,n\}$ for $n = 0$ as $\emptyset$ then the formula for a_n is also true for $n = 0$, for the empty set has only itself as subset, hence $a_0 = 1$. So one could start the recursion at $n = 1$.

Review of Problem 2.1

We have used the general techniques from Chapter 1 (examples, table etc.) again. In addition, we have **searched for a recurrence relation.**

We also encountered an important technique for counting problems: **We divided the objects that we wanted to count into classes.** Each of these classes for the counting problem of size n could be related to the same counting problem of size $n-1$, and this led us to the recursion.

In order to find these classes we focussed on what distinguishes the size n problem from the size $n-1$ problem: the element n of $\{1,\ldots,n\}$. Observing how this element matters in the problem – it can be contained in a subset or not – we found the solution.

In Chapter 5 you will get to know another way to tackle the problem of counting subsets.

2.3 Tilings with dominoes

Problem 2.2

In how many ways can you tile a rectangle of size $2 \times n$ with dominoes of size 1×2?

Here 'tiling' the rectangle means covering it with dominoes so that no two dominoes overlap and no domino extends outside the rectangle.

Investigation

▷ To **get a feel for the problem**, sketch a few **examples** and try to discover a rule. Squared paper is useful.

Think about it!

Figure 2.1 shows all possible tilings in the cases $n = 1,2,3,4$. We make a table for the number of tilings a_n for each n. Check foryourself that $a_5 = 8$.

n	1	2	3	4	5
a_n	1	2	3	5	8

Do you see a **pattern**? – How about this one: Each number in the lower row is the sum of its two predecessors. That is, $a_n = a_{n-1} + a_{n-2}$.

Does it go on like this? If yes, why?

▷ Based on the table we **conjecture** that the recursion $a_n = a_{n-1} + a_{n-2}$ holds for all n. How could we prove this? How can we relate the tilings of the $2 \times n$ rectangle (corresponding to a_n) to tilings of the $2 \times (n-1)$ and $2 \times (n-2)$ rectangles (corresponding to a_{n-1} and a_{n-2})?

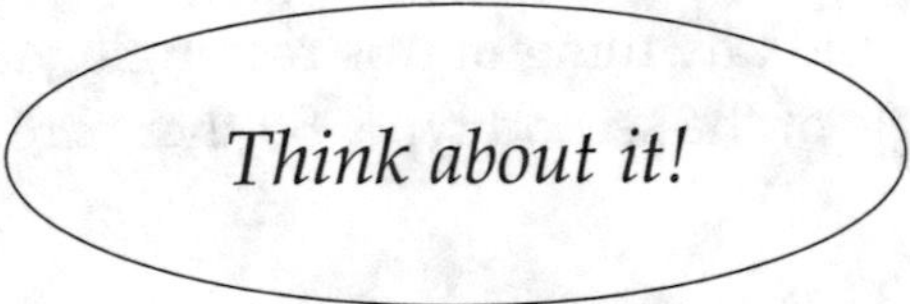

▷ What distinguishes a $2 \times n$ rectangle from a smaller rectangle? Something is added at the end. So let us look at the right-hand end of the $2 \times n$ rectangle, say. How can it be tiled? – Either there is one vertical domino, or two horizontal ones, see Figure 2.2. Every

$n = 1$

$n = 2$

$n = 3$

$n = 4$

Figure 2.1 Tilings of $2 \times n$ rectangles with dominoes

tiling is one of these two types. So we have **divided** the counting problem **into two classes** again. Let us count the tilings of each type separately.

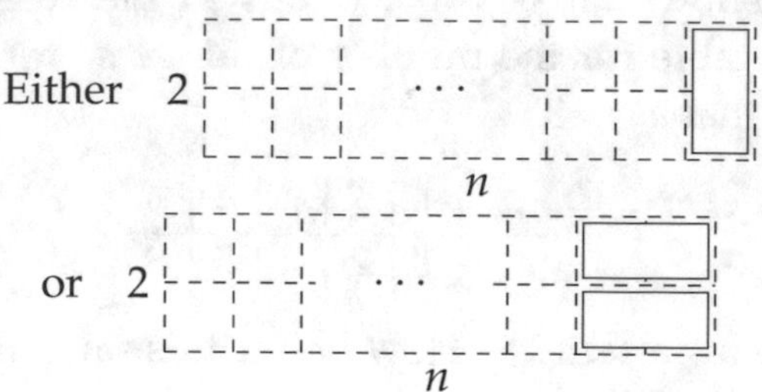

Figure 2.2 Two kinds of tilings

▷ How many tilings of the first type are there? To the left of the vertical domino there is a $2 \times (n-1)$ rectangle, and any tiling of this rectangle yields a tiling of the $2 \times n$ rectangle of the first type. So there are a_{n-1} tilings of the first type.

▷ Similarly, for a tiling of the second type we are left with a $2 \times (n-2)$-rectangle, and any tiling of this rectangle yields a tiling of the $2 \times n$ rectangle of the second type. So there are a_{n-2} tilings of the second type.

▷ Therefore, we get $a_n = a_{n-1} + a_{n-2}$.

▷ If you are not convinced, check the argument step by step for $n = 4$.

We have obtained the following solution.

! Solution of Problem 2.2 (derivation of a recurrence relation)

Let a_n be the number of domino tilings of a $2 \times n$ rectangle.

Claim: The recurrence relation $a_n = a_{n-1} + a_{n-2}$ holds for all $n \geq 3$.

Proof.
Every tiling of a $2 \times n$ rectangle is of one of the following two types (see Figure 2.2):

1. There is a vertical domino at the right-hand end.
2. There are two horizontal dominoes at the right-hand end.

For any type-1 tiling, a $2 \times (n-1)$ rectangle remains to the left of the vertical domino. Let us call the $2 \times n$ rectangle R and the

$2 \times (n-1)$ rectangle R'. The type-1 tilings of R correspond precisely to all the tilings of R': Any tiling of R' yields a type-1 tiling of R, by simply adding the vertical domino at the right-hand end. Conversely, removing the right-most vertical domino from a type-1 tiling of R yields a tiling of R'.

Therefore, R has a_{n-1} type-1 tilings.

Similarly, for any type-2 tiling of R a $2 \times (n-2)$ rectangle R'' remains, and type-2 tilings of R correspond precisely to tilings of R''. Therefore, R has a_{n-2} type-2 tilings.

Altogether there are $a_{n-1} + a_{n-2}$ tilings of the $2 \times n$ rectangle, so $a_n = a_{n-1} + a_{n-2}$. q. e. d.

We have proved the recurrence relation. We will solve it in the next section. !

Review of Problem 2.2

Using a table we arrived at a conjecture for a recurrence relation. As in the previous problem we were able to prove the recurrence relation by dividing the objects to be counted (the tilings) into **two classes** (corresponding to type-1 and type-2 tilings), and relating each class to a smaller problem of the same kind.

With a little more experience we might have skipped the 'conjecture' part, and tried to find a recursion directly. The question "How can I reduce the problem of size n to smaller problems of the same kind?" and the focus on one end of the strip might have led us to the right idea. This kind of thinking will be needed for problems where it is hard to discover a pattern. We will discuss such a problem later, see Problem 2.4.

What did we accomplish? The recurrence relation allows us to extend the table quickly and compute, for example, a_{10}, see Table 2.1. To find a_{10} without the recurrence relation would be very tiresome! The a_n are called FIBONACCI *numbers*.[3]

[3]Sometimes the sequence of numbers with index shifted by one are called FIBONACCI numbers: $f_1 = 1, f_2 = 1, f_3 = 2, f_4 = 3$, in general $f_n = a_{n-1}$.

n	1	2	3	4	5	6	7	8	9	$10\cdots$
a_n	1	2	3	5	8	13	21	34	55	$89\cdots$

Table 2.1 The FIBONACCI numbers

2.4 Solving the FIBONACCI recurrence relation

We saw that the recurrence relation

$$a_n = a_{n-1} + a_{n-2} \tag{R}$$

allows us to compute a_n easily. However, we still need to do about n additions to obtain a_n, and for large n this is tiresome. We now want to solve the recurrence relation, that is, we want to find a formula that allows us to compute a_n directly from n.

The initial conditions are $a_1 = 1, a_2 = 2$. In this case we need two initial condition since each a_n is determined by the two previous terms in the sequence. The calculations below are simplified if we define $a_0 = 1$, then (R) also holds for $n = 2$.

? Problem 2.3

Find a closed formula for the sequence $a_0, a_1, \ldots$ which is defined by the recurrence relation (R) (valid for $n \geq 2$) and the initial conditions

$$a_0 = 1, \quad a_1 = 1. \tag{IC}$$

A **closed formula** is an expression that allows us to compute a_n directly from n. For example, the solution of the recurrence relation $a_n = 2a_{n-1}$, $a_0 = 1$ is the closed formula $a_n = 2^n$.

Investigation and solution

▷ Can you guess a formula from Table 2.1? Does any of the usual suspects work, for example n, n^2, 2^n, $n!$, or a combination (sum, product, ...) of those? Try it!

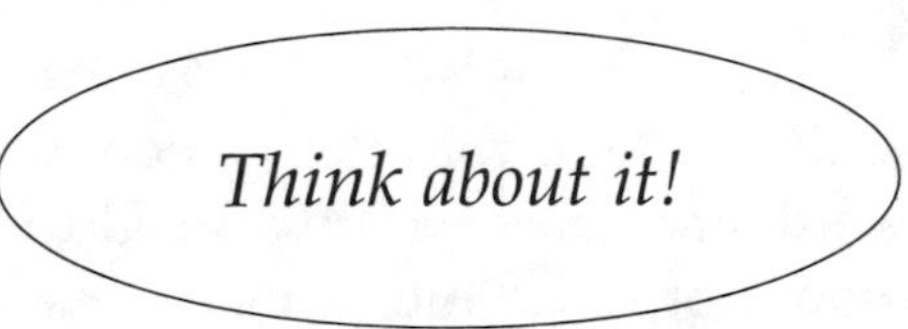

▷ You didn't find a formula? Don't worry, to guess the formula is quite impossible (as you will agree when you see it). However, there is an ingenious method to find a formula. Rather than just presenting how it works, I will develop it together with you from simple principles.

▷ Let us use our problem-solving strategies: The problem is too difficult. **Simplify!**

How? We need to satisfy two conditions: (R) and (IC). Let us first forget about (IC) and focus on (R) only.

▷ So the question is: Can we find *any* expression in n which satisfies the recursion (R) for all n? Let us play a little with **examples**: Try $a_n = n$: Indeed, $a_3 = a_2 + a_1$ is true since $3 = 2 + 1$, but $4 \neq 3 + 2$ shows that (R) is not true for $n = 4$. So $a_n = n$ does not work. Maybe this was too special. Let us try more general expressions, for example $a_n = rn + s$ (for fixed $r, s \in \mathbb{R}$)[4], n^2, n^k (for fixed $k \in \mathbb{N}$), ... You will find that none of this works. Let us try an exponential again, but instead of the special 2^n we leave the base undetermined. That is, we make the[5]

$$\textit{Ansatz:} \quad a_n = \alpha^n \quad \text{for some real number } \alpha \neq 0$$

We plug this into the recurrence relation $a_n = a_{n-1} + a_{n-2}$ and obtain

$$\alpha^n = \alpha^{n-1} + \alpha^{n-2}$$

or equivalently (divide by α^{n-2})

$$\alpha^2 = \alpha + 1. \tag{2.6}$$

[4]Let's do it: Plug $a_n = rn + s$ into the recursion; this yields $rn + s = r(n - 1) + s + r(n - 2) + s$. A little algebra yields $rn = 3r - s$. Can this be true for all n? Obviously not, unless $r = 0$ and $s = 0$. But then $a_n = 0$ for all n – indeed a solution but an uninteresting one.

[5]The German word Ansatz means attempt or approach. In this mathematical context it is also used in English.

This shows that $a_n = \alpha^n$ satisfies the recurrence relation (R) *for all* n if and only if α is a solution of equation (2.6). This is a big step forward since now we only need to find the solutions of one equation, not of infinitely many ($a_n = a_{n-1} + a_{n-2}$ for $n = 2, 3, 4, \ldots$).

Equation (2.6) is a quadratic equation for α which may be solved by the well-known formula. We rearrange to $\alpha^2 - \alpha - 1 = 0$ and obtain the solutions $\frac{1}{2} \pm \sqrt{\frac{1}{4} + 1}$ or

$$\alpha = \frac{1 \pm \sqrt{5}}{2}.$$

We introduce the notation

$$\beta = \frac{1 + \sqrt{5}}{2}, \quad \gamma = \frac{1 - \sqrt{5}}{2}$$

and call the corresponding sequences b_n, c_n. This will allow us to continue to use the letter a_n as in the beginning, for the sequence satisfying both (R) and (IC).

▷ What have we got? We have found two solutions for the recurrence relation (R):

$$b_n = \beta^n, \quad c_n = \gamma^n. \tag{2.7}$$

This is a success: we have attained our interim goal, finding a solution for the simplified problem. We even found two solutions, and this will be very useful.

Now we need to look at the initial conditions. Let us check whether one of our solutions satisfies (IC). Using $\beta^0 = \gamma^0 = 1$ we get

	n	0	1	…
known:	b_n	1	β	…
known:	c_n	1	γ	…
wanted:	a_n	1	1	…

(2.8)

We see that b_n, c_n satisfy the initial condition for $n = 0$ but not for $n = 1$.

▷ We need another idea. Did we find *all* solutions of (R)? Or can we produce more solutions from the two that we found?

▷ **New from old!** There are two simple ways to produce new solutions from old ones:

1. Multiply by a constant: If $b_0, b_1, \ldots$ solves (R) and B is any real number then $Bb_0, Bb_1, \ldots$ also solves (R), because

$$b_n = b_{n-1} + b_{n-2} \Rightarrow Bb_n = Bb_{n-1} + Bb_{n-2}$$

for all $n \geq 2$.

2. Add: If $b_0, b_1, \ldots$ and $c_0, c_1, \ldots$ both solve (R) then so does $d_0, d_1, \ldots$ defined by $d_n = b_n + c_n$, because we can simply add the equations:

$$\begin{array}{rrcrcr} & b_n & = & b_{n-1} & + & b_{n-2} \\ + & c_n & = & c_{n-1} & + & c_{n-2} \\ \hline & b_n + c_n & = & (b_{n-1} + c_{n-1}) & + & (b_{n-2} + c_{n-2}) \end{array}$$

so $d_n = d_{n-1} + d_{n-2}$ for all $n \geq 2$.

We can also combine both ideas: First apply 1. to the b_n and the c_n, with two constants B and C, then add the results. This is called the[6]

> **Superposition principle:** If $b_0, b_1, \ldots$ and $c_0, c_1, \ldots$ solve (R) and B, C are any real numbers then $a_0, a_1, \ldots$ defined by
>
> $$a_n = Bb_n + Cc_n \quad \text{for all } n$$
>
> also solves (R).

The point is that since B and C are arbitrary, we get quite many solutions of (R), so there is a chance that among them there is one which also satisfies (IC).

▷ So now we need to find the numbers B, C such that these a_n also satisfy (IC)[7]. We write out the equation $a_n = Bb_n + Cc_n$ for $n = 0$ and $n = 1$ and plug in the values from the table (2.8):

$$\begin{array}{lcclcl} n = 0: & 1 & = & B & + & C \\ n = 1: & 1 & = & B\beta & + & C\gamma \end{array}$$

[6]The word originates from physics where it describes, for example, the superposition of two waves to form a total wave. The superposition principle is always valid for linear equations.

[7]Before jumping into calculations let us think why this might work: We have to satisfy two initial conditions (equations), and we have to determine two numbers B, C (unknowns). So the number of equations is the same as the number of unknowns. Rule of thumb: usually this will work.

This is a linear system of equations for B, C and can be solved easily: In order to eliminate one of the unknowns, say B, we multiply the first equation by β and subtract the second equation.

$$\begin{array}{rcrcr} \beta & = & B\beta & + & C\beta \\ - \quad 1 & = & B\beta & + & C\gamma \\ \hline \beta - 1 & = & 0 & + & C(\beta-\gamma) \end{array}$$

We obtain

$$C = \frac{\beta - 1}{\beta - \gamma} \quad \text{and then } B = -\frac{\gamma - 1}{\beta - \gamma},$$

where the second equation follows from the first using $1 = B + C$ after a short calculation.

▷ We now use the values of B, C and b_n, c_n in the formula $a_n = Bb_n + Cc_n$ and obtain: The sequence of numbers

$$a_n = -\frac{\gamma - 1}{\beta - \gamma}\beta^n + \frac{\beta - 1}{\beta - \gamma}\gamma^n,$$

satisfies both the initial condition (IC) and the recursion (R). Since (IC) and (R) uniquely determine the sequence, this must be our solution.

We can write this a little more neatly: From $\beta = \frac{1+\sqrt{5}}{2}$, $\gamma = \frac{1-\sqrt{5}}{2}$ we get $\beta - 1 = -\gamma$, $\gamma - 1 = -\beta$ and $\beta - \gamma = \sqrt{5}$, so

$$\boxed{a_n = \frac{1}{\sqrt{5}}\left[\left(\frac{1+\sqrt{5}}{2}\right)^{n+1} - \left(\frac{1-\sqrt{5}}{2}\right)^{n+1}\right]} \tag{2.9}$$

This is the desired closed formula!

Let us summarize our method:

Method: Solving the FIBONACCI recurrence relation

In order to solve the recurrence relation $a_n = a_{n-1} + a_{n-2}$ with initial condition $a_0 = 1$, $a_1 = 1$, proceed as follows:

1. Make the ansatz $a_n = \alpha^n$. This leads to the equation $\alpha^2 = \alpha + 1$. Denote its solutions by β, γ. Then β^n, γ^n are solutions of the recurrence relation.
2. Write $a_n = B\beta^n + C\gamma^n$, plug in the initial conditions and solve

the resulting linear system of equations for B, C. This yields the desired formula for a_n.

Further remarks: Solving the FIBONACCI recurrence relation

- Is the solution (2.9) *reasonable*? At first glance it does not seem so: Clearly all a_n must be integers since the initial values are and the recurrence relation then always produces integers. However, the formula contains the irrational number $\sqrt{5}$. For example, $\frac{1+\sqrt{5}}{2} = 1{,}618\ldots$. See Exercise E 5.16 for the resolution of this apparent paradox.
- Let us state again *why* the method works: The right hand side of formula (2.9) satisfies by construction both the initial condition (IC) and the recurrence relation (R). Since the left hand side also satisfies (IC) and (R), and since (IC) and (R) determine the sequence uniquely, the formula must be correct for all n.
- This proves the correctness of formula (2.9). You could provide an additional proof using mathematical induction. You may want to do this for exercise. But logically it is not needed.
- *(Generalization)* The same method works also for other initial conditions, and even for other recurrence relations. More precisely, it works for any recurrence relation

$$a_n = pa_{n-1} + qa_{n-2} \tag{2.10}$$

where p, q are real numbers, with any two given initial values for a_0, a_1. These are called *linear recurrence relations with constant coefficients.*
However, for some values of p, q the method needs to be modified slightly, see the exercises.
The method also works for linear recurrence relations with constant coefficients and length greater than two, for example, $a_n = a_{n-1} - a_{n-2} + a_{n-3}$. But now instead of the quadratic equation for α you need to solve an equation of higher degree (in the example $\alpha^3 - \alpha^2 + \alpha - 1 = 0$), which may be difficult. Of course one needs to give three initial values.

- An important part of the method is the superposition principle. This is a special case of the very general idea of *linear combination*, a fundamental notion in linear algebra.

  If you know some linear algebra then you can understand the method as follows: In the vector space of all sequences $a_0, a_1, \ldots$ the subset of those sequences that satisfy the recurrence relation is a linear subspace. This subspace is two-dimensional since any one of these sequences is determined by two initial conditions. The special sequences $\beta^0, \beta^1, \ldots$ and $\gamma^0, \gamma^1, \ldots$ lie in this subspace and are linearly independent, as follows easily from $\beta \neq \gamma$. Therefore they form a basis of the subspace, therefore the sequence $a_0, a_1, \ldots$ that we are looking for must be representable as some linear combination of them.

- There is another elegant method for solving linear recurrence relations with constant coefficients, using so-called generating functions, see (Aigner, 2007).

By the way ...

To find the number of domino tilings for arbitrary rectangles is much harder than in the case of $2 \times n$ rectangles (see also Exercise E 2.12). In 1961, both Kasteleyn and (independently) Temperley and Fisher found a remarkable general formula: The number of domino tilings of an $m \times n$ rectangle is

$$\prod_{j=1}^{m} \prod_{k=1}^{n} \left(4\cos^2 \frac{\pi j}{m+1} + 4\cos^2 \frac{\pi k}{n+1} \right)^{1/4}.$$

The two big Pi's are short notation for 'product'. They tell us to plug in all possible values of j and k in the parentheses and multiply all the results. For example, if $n = 2$, $m = 3$ then this is a product of 6 factors.

This formula is very mysterious: Why do cosines appear? Why is the product an integer? The reason why Kasteleyn and Temperley-Fisher were interested in this problem is that it appears in theoretical physics.[8]

[8] You can find more on this, and a guide to the literature, on the Wikipedia page on domino tilings.

2.5 Triangulations

We now consider a counting problem where you cannot easily find a pattern or recurrence relation by simply staring at a few numbers. But we will still be able to solve it by looking systematically for a recurrence relation.

? Problem 2.4

Let $n \geq 3$ and P be a convex n-gon.[9] *A* triangulation *of P is a subdivision of P into triangles, using non-intersecting lines connecting vertices of P, see Figure 2.3. How many triangulations does P have? Find a recurrence relation.*

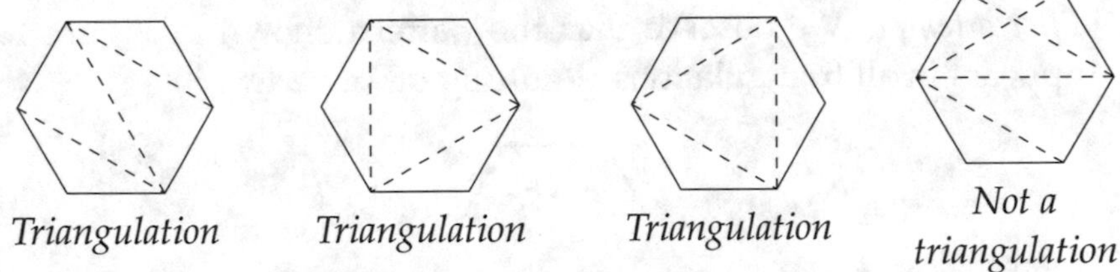

Figure 2.3 What is a triangulation? Examples for $n = 6$

We will consider two triangulations as different even if we can obtain one from the other by a rotation or reflection. For example, the three triangulations in Figure 2.3 are all different.

Investigation of Problem 2.4

▷ We are looking for the number of triangulations of a convex n-gon. Let us call this number T_n.

▷ By trying all possibilities systematically (do it!) we find:

n	3	4	5	6
T_n	1	2	5	14

[9]That is, a triangle, quadrilateral, pentagon, hexagon if $n = 3, 4, 5, 6$. For n-gon we also say **polygon**. A polygon is **convex** if all lines connecting two of its vertices lie in the polygon.

There is no obvious pattern in the sequence $1, 2, 5, 14$. So we try to **look for a recurrence relation directly from the problem**. How can we obtain triangulations of an n-gon from triangulations of polygons with fewer vertices? Can we divide all triangulations of the n-gon into classes which correspond to triangulations of smaller polygons?

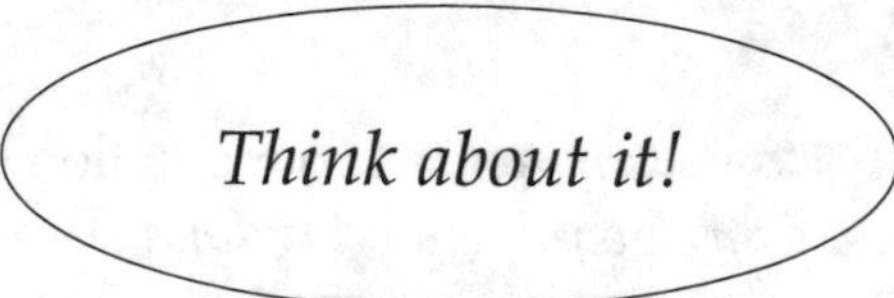

▷ There are many ways to approach this. Maybe you had one of the following ideas.
First attempt: We observe that the pattern shown in Figure 2.4 appears in all triangulations – 'cutting off' a vertex.

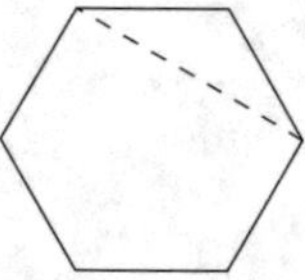

Figure 2.4 6-gon with cut-off vertex

After cutting off the vertex we are left with a polygon that has $n-1$ vertices, which we can triangulate in T_{n-1} ways. Since we can cut off any of the n vertices and then, in each case, have T_{n-1} ways to triangulate the rest, we conjecture the recurrence relation

$$\mathrm{T}_n \overset{?}{=} n \cdot \mathrm{T}_{n-1} \tag{2.11}$$

So here we tried to divide all the triangulations of the n-gon into n classes, corresponding to the cut-off vertex.

Looking at the table we see that the recursion is wrong. But why? Where was the mistake in our argument?

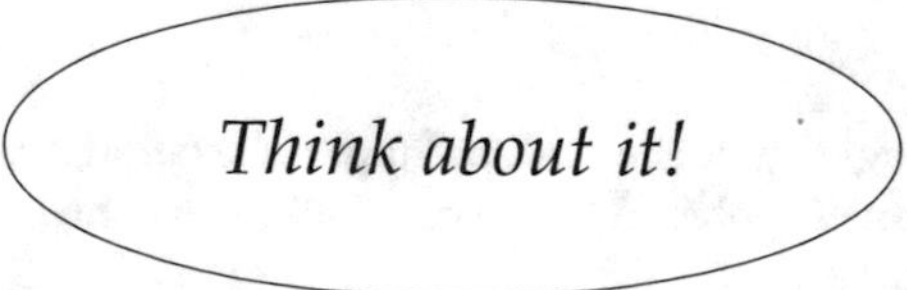

In our argument we counted some triangulations multiple times. For example, the triangulation in Figure 2.5 is counted twice. It lies in two of the classes. How can we remedy this? We could try

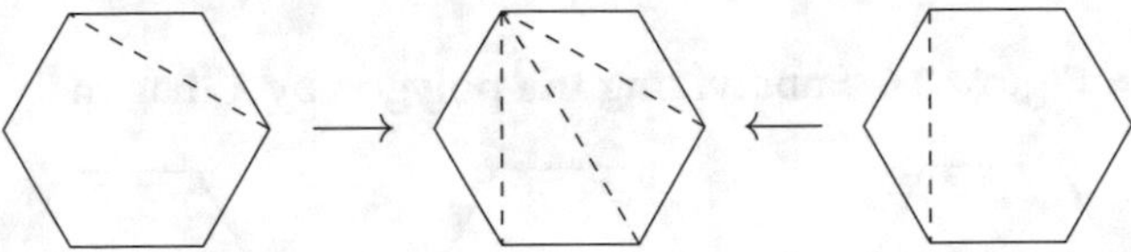

Figure 2.5 A triangulation which is counted twice

to subtract a suitable number from $n\,\mathrm{T}_{n-1}$ corresponding to such multiple counts. This is possible but a little tricky, so instead we will follow another path.[10] Anyway, we see that *conjecture* (2.11) *is wrong*.

By the way: When proceeding like this we would still have to prove what we observed in examples: That every triangulation has at least one cut-off vertex. This is not hard, for example using induction or using the extremal principle, see Exercise E 10.9.

▷ *Second attempt:* Let us look at Figure 2.3 again and draw some more triangulations. Let us focus on one vertex, say the right-hand vertex of the hexagon. We see that there is always a line (diagonal) from the right vertex to some other vertex, or else the right vertex is cut off.

Such a diagonal allows us to reduce the problem to smaller problems of the same kind since it cuts the hexagon into two smaller polygons, see Figure 2.6. In this example, there are two quadrangles, having T_4 triangulations each. Since we can combine any two of them and obtain a triangulation of the hexagon, we get $\mathrm{T}_4 \cdot \mathrm{T}_4$ triangulations. In addition, we get triangulations arising from other ways to draw the diagonal, and we can count those in a similar way.

But again we have the problem of multiple counts, see Figure 2.7.

▷ *Third attempt:* We have seen that we need to be careful: When

[10] If you want to know more about how to deal with such multiple counts systematically, look up 'Inclusion-exclusion principle' in a book on combinatorics or on Wikipedia. Then try to use this principle here.

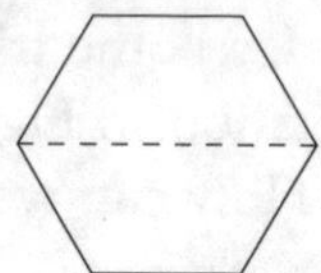

Figure 2.6 Subdividing the polygon by a diagonal

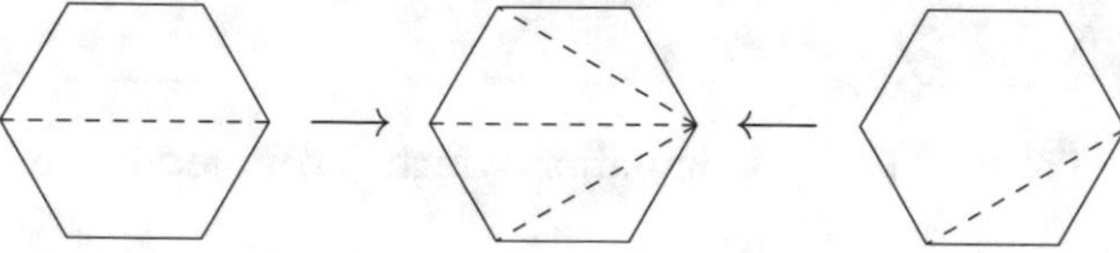

Figure 2.7 A triangulation which is counted twice in the second attempt

dividing the problem into classes then the classes must be *disjoint*.[11] **We need a *disjoint* subdivision of the problem.**

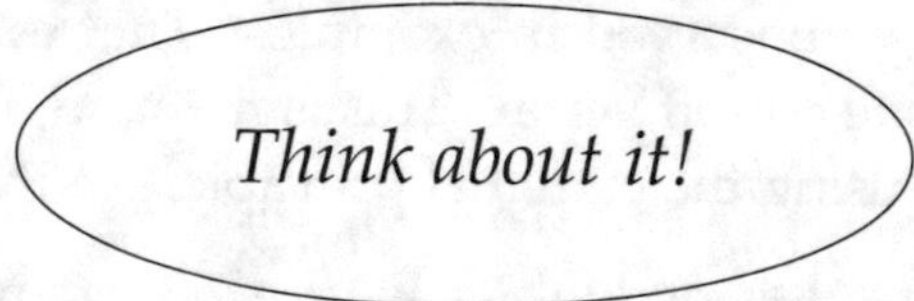

Here is a new idea: We focus on one side of the n-gon, say the bottom side in Figure 2.8. Any triangulation must have precisely one triangle which contains this side. So we can classify the triangulations according to which triangle this is. For $n = 6$ there are 4 possibilities for this triangle as in the figure, therefore there are 4 classes. Note that the triangle is determined by the location of its third vertex, and this can be any vertex except the two bottom ones; this explains why there are $6 - 2 = 4$ possibilities.

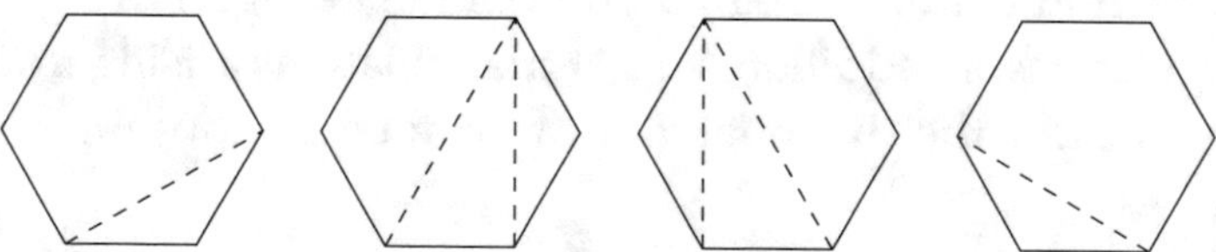

Figure 2.8 Classes into which we subdivide our problem

This is a disjoint subdivision since in any triangulation there can

[11]Two sets (or classes, which is the same) are called disjoint if they have no common elements. Here this means that each triangulation must lie in precisely *one* class.

only be one triangle containing the bottom side. Compare this with the second attempt: There can be several diagonals leaving the right-hand vertex, and this led to the double count there, i.e. non-disjoint classes.

To the left and the right of the dashed triangle we get convex polygons with fewer than 6 vertices. By counting their triangulations and combining them we can count the triangulations of the hexagon.

We now carry out this idea for general n.

Solution of Problem 2.4

Consider an arbitrary convex n-gon and number its vertices $P_1, P_2, \ldots, P_n$ as in Figure 2.9. We divide the triangulations of the n-gon into classes, where each class is determined by the third vertex of that triangle in the triangulation which contains the side P_1P_n.

Call this triangle D and its third vertex P_k. Here k can take any of the values $2, \ldots, n-1$, see Figure 2.9. So we get one class for each such value of k.

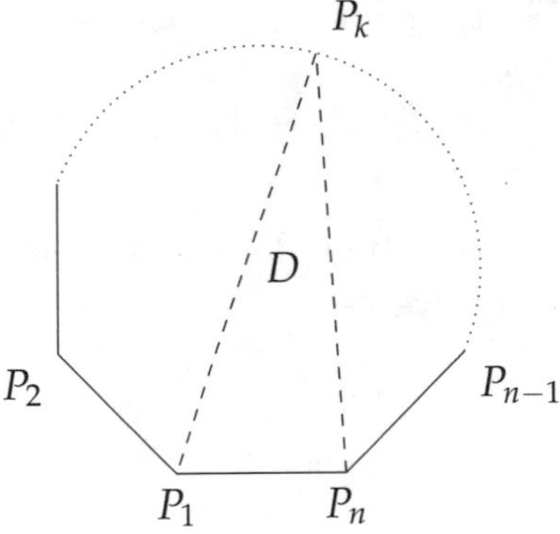

Figure 2.9 One of the classes into which we subdivide the problem

Let us count how many triangulations are in the class labelled by k. To the left of D we have a k-gon and to the right of D we have an $(n-k+1)$-gon (check this!). We can triangulate each of these polygons and obtain a triangulation of the n-gon. Since we can combine every left triangulation with every right triangulation, we multiply their numbers. This gives us the number of triangulations in the class k.

The cases $k = 2$ and $k = n - 1$ are special since there is no actual polygon on the left or right, respectively, only a '2-gon' which cannot be triangulated.

	left of D		right of D		triang.
k	vertices	triang.	vertices	triang.	combined
2	no polygon	–	$n-1$	T_{n-1}	T_{n-1}
3	3	T_3	$n-2$	T_{n-2}	$T_3 \cdot T_{n-2}$
4	4	T_4	$n-3$	T_{n-3}	$T_4 \cdot T_{n-3}$
$\vdots$	$\vdots$	$\vdots$	$\vdots$	$\vdots$	$\vdots$
k	k	T_k	$n-k+1$	T_{n-k+1}	$T_k \cdot T_{n-k+1}$
$\vdots$	$\vdots$	$\vdots$	$\vdots$	$\vdots$	$\vdots$
$n-2$	$n-2$	T_{n-2}	3	T_3	$T_{n-2} \cdot T_3$
$n-1$	$n-1$	T_{n-1}	no polygon	–	T_{n-1}

Thus we have found the number of triangulations in each class. Since the classes are disjoint, the total number of triangulations of the n-gon is their sum. We get

$$\begin{aligned} T_n &= T_{n-1} + T_3 \cdot T_{n-2} + T_4 \cdot T_{n-3} + \cdots + T_{n-2} \cdot T_3 + T_{n-1} \\ &= T_{n-1} + \sum_{k=3}^{n-2} T_k \cdot T_{n-k+1} + T_{n-1} \,. \end{aligned}$$

See the list of symbols for an explanation of the summation sign $\sum$. The formula simplifies if we set $T_2 := 1$:

$$T_n = \sum_{k=2}^{n-1} T_k \cdot T_{n-k+1} \tag{2.12}$$

This is the recurrence relation we were looking for. It allows us to extend the table for the T_n easily. !

Review of Problem 2.4

We were not able to detect a pattern from our table. Therefore we tried systematically to find a recursion, i.e. a method of reducing the problem to a smaller problem of the same kind.

The first two ideas for such a reduction (cutting off a vertex, considering diagonals from a fixed vertex) did not work since they led to multiple counts.

We finally found a recurrence relation by shifting focus from vertices to sides, which allowed us to **subdivide the problem into disjoint classes** and express the number of triangulations in each class using triangulation numbers of smaller polygons. ○

Remarks

- The numbers $C_n = T_{n+2}$ are called *Catalan numbers*. They also occur in many other counting problems. For example, C_{n+1} is the number of bracketings of n symbols, see Exercise E 2.11. See Wikipedia for more examples. The Catalan numbers satisfy the (slightly prettier) recursion

$$C_{n+1} = \sum_{k=0}^{n} C_k C_{n-k}$$

which you can easily derive from (2.12) (exercise!).

- To solve the recurrence relation for the T_n (or the C_n) is harder than for the FIBONACCI numbers. The main difference is that this recurrence relation is non-linear while the FIBONACCI recurrence relation is linear[12]. A very elegant method which works is generating functions, see (Aigner, 2007), for example.

2.6 Toolbox

In this chapter you learned a few special strategies which are useful for problems of the kind "For all n find the number a_n of objects or configurations of a certain kind":

- **Subdivide into classes:** Subdivide the set of configurations that you want to count into classes which are easier to count. The classes must be disjoint.

[12]This is a general phenomenon in mathematics: non-linear problems are often harder than linear problems.

- ❑ **Recursion:** If counting the classes involves the same type of counting problem but for a smaller problem size, you get a recurrence relation.

  A recurrence relationallows you to calculate a_n quickly even for values of n where counting the objects/configurations directly would be difficult.

These strategies are special cases of the general strategies **simplify** and **interim goals**.

You also learned a technique for solving linear recurrence relations with constant coefficients.

Exercises

Exercises E 2.1 to E 2.4 serve to deepen your understanding of the method of solving the FIBONACCI recurrence relation.

2 E 2.1 Let a_n be the FIBONACCI numbers and define β, γ as in (2.7).

a) Show that a_n equals the integer nearest to $\frac{\beta^{n+1}}{\sqrt{5}}$.

b) (Using a little bit of calculus) Show that $\frac{a_{n+1}}{a_n}$ converges for $n \to \infty$ to the golden ratio $\beta = \frac{1+\sqrt{5}}{2}$.

1-2 E 2.2 Show that the method for solving the Fibonacci recurrence relation works for all recurrence relations (2.10) for which the equation $\alpha^2 - p\alpha - q = 0$ has two *different* solutions. Solve the recurrence relation $a_n = 2a_{n-1} + 3a_{n-2}$ with the initial condition $a_0 = 1, a_1 = 3$ and with the initial condition $a_0 = 1, a_1 = 2$.

2-3 E 2.3 The solutions of $\alpha^2 - p\alpha - q = 0$ can also be complex numbers. This is a place where complex numbers are really useful. If you know about them, consider the recurrence relation $a_n = \sqrt{2}a_{n-1} - a_{n-2}$, $a_0 = 0, a_1 = 1$, calculate the first ten terms, observe a pattern and then explain the pattern using the fact that the numbers $\alpha = \frac{1}{\sqrt{2}}(1 \pm i)$ satisfy $\alpha^8 = 1$.[13] Even more interesting is the recurrence relation $a_n = \frac{1}{2}a_{n-1} - a_{n-2}$, $a_0 = 0$, $a_1 = 1$. Calculate a few terms. The

[13] You can check this by direct calculation. But it is nicer to prove it geometrically using the geometric meaning of complex numbers and their multiplication.

sequence does not grow unboundedly, but it is also not periodic. Can you explain this? How is it related to the fact that the number $\frac{1}{\pi}\arccos(\frac{1}{4})$ is irrational (and how can you prove that it is?)?

E 2.4 What do you do if $\alpha^2 - p\alpha - q = 0$ has only one solution α? 2-3
Investigate!

For example, if $p = 2, q = -1$ then the only solution is $\alpha = 1$, so the recursion $a_n = 2a_{n-1} - a_{n-2}$ has the solution $a_n = 1^n = 1$ for all n. Find a second solution which is independent of the first (i.e. where not all a_n are equal). Find the general solution (with any initial conditions) for this recurrence relation. Make a conjecture how to proceed in general when there is only one solution α. Prove your conjecture.

E 2.5 Solve the recurrence relations $a_n = a_{n-1} + 1$, $a_n = a_{n-1}^2$ and 1
$a_n = na_{n-1}$ (for all $n \geq 2$), with initial condition $a_1 = 1$.

E 2.6 Fold a paper strip n times, halving its length every time. Then 1
unfold again. How many folding edges do you get?

E 2.7 Suppose that n people meet and shake hands, every person 1-2
exactly once with every other person. That's how many handshakes? Solve the problem recursively.

E 2.8 What is the number of trains of length n whose coaches each 2
have length 1 or 2? How many of these trains are symmetric, that is, they look the same from front and back? Here is a symmetric and an asymmetric train of length 4:

E 2.9 Determine the number of sequences of zeros and ones of length 2
n which contain no consecutive ones. For example for $n = 3$ there are 5 such sequences: 000, 001, 010, 100, 101.

E 2.10 How many subsets of $\{1, \ldots, n\}$ do not contain any pair of 2
consecutive numbers?

E 2.11 Find a recurrence relation for the number of possible bracket- 2-3
ings of n factors $abc\cdots$. Here a bracketing must be such that when evaluating the expression only two factors are multiplied in each

step. For example, the possible bracketings of $n = 4$ factors are $(ab)(cd)$, $a(b(cd))$, $a((bc)d)$, $((ab)c)d$, $a((bc)d)$, but $(abc)d$ would not be a bracketing.

3 E 2.12 In how many ways can you tile a rectangle of size $3 \times n$ with dominoes of size 1×2?

3 E 2.13 How many natural numbers are there that have n digits, where only the digits 1, 2 and 3 occur and where consecutive digits differ by at most one?

3 E 2.14 Nim is a game for two players. At the start, n matches are on the table. Players take turns, and in each move a player can take 1, 2 or 3 matches. The player who takes the last match wins. For which n can the first player force victory, that is play in such a way that he/she wins, whatever the second player does?

2 E 2.15 For $n \in \mathbb{N}$ denote by o_n the number of ways to write n as ordered sum of *odd* natural numbers. Count n itself (one summand) as one of these ways if n is odd. For example

$$3 = 1 + 1 + 1 \text{ and } 4 = 1 + 1 + 1 = 1 + 3 = 3 + 1$$

so $o_3 = 2$ and $o_4 = 3$. Find o_n for $n = 1, 2, 3, 4, 5, 6$, state a conjecture and prove it. Compare Exercise E 1.9.

3 Mathematical induction

Mathematical induction is one of the most important methods for proving statements of the form "For all natural numbers ...". At its heart it is another instance of the idea of recursion: reduce the problem to a smaller problem of the same kind. Mathematical induction implements this idea for proofs, while recurrence relations are used in problems where you want to determine some quantity.

In this chapter we consider two problems where mathematical induction can be used, with more examples to come in the following chapters. We also discuss the limitations of induction. Along the way you will encounter another important element of problem-solving and any scientific work: introducing suitable terms or concepts.

3.1 The induction principle

Mathematical induction is based on the following principle:

Induction principle

Let $A(n)$ be a statement about natural numbers n. Suppose

1. **Base case (BC):** $A(1)$ is true.
2. **Inductive step (IS):** For all $n \in \mathbb{N}$ the following implication is true:

   If $A(n)$ is true, then $A(n+1)$ is also true.

Then $A(n)$ is true for all $n \in \mathbb{N}$.

In the inductive step $A(n)$ is called the **inductive hypothesis** and $A(n+1)$ the **inductive claim.**

The induction principle can be visualized as the *domino effect*: put a row of dominoes upright on the table in such a way that each one will knock over the next one when falling. Now if you knock over the first one then they will all fall, see Figure 3.1.

D. Grieser, *Exploring Mathematics*, Springer Undergraduate Mathematics Series, https://doi.org/10.1007/978-3-319-90321-7_3

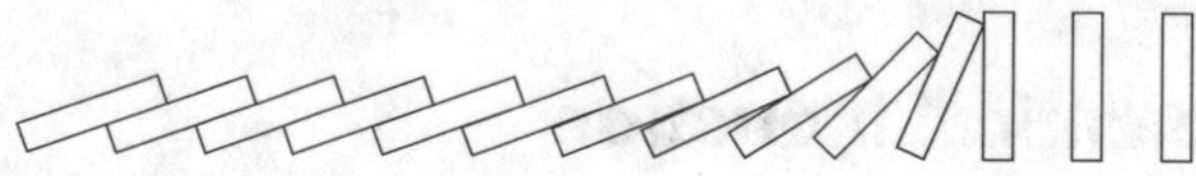

Figure 3.1 Domino effect

Here is a simple inductive proof for illustration:

Claim: For all $n \in \mathbb{N}$ we have[1]

$$\sum_{k=1}^{n} k = \frac{n(n+1)}{2}.$$

Proof.
We use mathematical induction. Here $A(n)$ is the statement that the formula holds for the value n.
Base case: The statement $A(1)$ is: $1 = 1\cdot 2/2$. This is obviously true.
Inductive hypothesis: Let n be arbitrary, and assume $A(n)$ is true, so $\sum_{k=1}^{n} k = n(n+1)/2$ holds.
Inductive claim: $A(n+1)$ is true, that is $\sum_{k=1}^{n+1} k = (n+1)(n+2)/2$.
Proof (inductive step): We use $A(n)$ and compute:

$$\begin{aligned}\sum_{k=1}^{n+1} k &= \left(\sum_{k=1}^{n} k\right) + (n+1) \\ &= \frac{n(n+1)}{2} + (n+1) = \left(\frac{n}{2} + 1\right)(n+1) = \frac{n+2}{2}(n+1) \\ &= \frac{(n+1)(n+2)}{2},\end{aligned}$$

which was to be shown. q. e. d.

Practical note. After a while the correct use of mathematical induction will become second nature to you. Until then it is useful to write out the inductive hypothesis and inductive claim fully, in order to avoid mistakes.

Often we use **variants of the induction principle**. Here are some variations, which can also be combined. You will find examples in this and the next chapters.

- **IS** replaced by: For all $n \geq 2$ we have $A(n-1) \Rightarrow A(n)$.

[1] From now on we use the summation notation. See the list of symbols.

- **BC** for $n = 0$, **IS** for all $n \geq 0$. More generally: **BC** for $n = n_0$, **IS** for all $n \geq n_0$, where n_0 is some natural number.[2]

- **IS** replaced by: For all $n \in \mathbb{N}$ we have

$$A(1), \ldots, A(n) \text{ imply } A(n+1).$$

- **BC:** $A(1)$ and $A(2)$ are true.
  IS: For all $n \in \mathbb{N}$ we have: $A(n), A(n+1)$ imply $A(n+2)$.

What induction cannot do: Mathematical induction (or induction for short[3]) is a useful method of proof, but it has its limitations: It allows us to *prove* the formula for $\sum_{k=1}^{n} k$, *once we conjecture it*. It does not help us to *find* the formula. If you are faced with the task of finding a formula for the sum of squares, $\sum_{k=1}^{n} k^2$, you must come up with some idea. However, if someone tells you the formula $\sum_{k=1}^{n} k^2 = n(n+1)(2n+1)/6$ then you can prove it using mathematical induction.

Induction does not tell you "where the formula comes from"[4]. An inductive proof will give you security, but will often leave you unsatisfied that you haven't really understood the problem.

So induction is like a crutch. If you don't find a better argument then it is very useful. But it's better to do without it whenever possible.

This limitation is also reflected in the **relation of mathematical induction and recursion**: When you have derived a recurrence relation for numbers a_n and guessed a closed formula for the a_n, then you can prove this formula using induction. The recurrence relation will be the core of the inductive step; the initial condition will be the base case. You saw an example of this in the formal solution of Problem 2.1. But induction will not help you *find* a closed formula for a_n.

---

[2]Of course this will prove $A(n)$ only for $n \geq n_0$.

[3]In mathematics, induction always means mathematical induction. This is not to be confused with inductive reasoning as it is used in most other sciences, which means to use a few special cases to conclude a general rule. This is appropriate for other sciences but is not a sufficient argument for a mathematical proof. See Chapter 7.2 for an example.

[4]For $\sum_{k=1}^{n} k$ the Gauss trick is a way to find the formula, for $\sum_{k=1}^{n} k^2$ see Problem 5.9 for a hint.

3.2 Colourings

In the example above we used mathematical induction to prove a formula. But it can also be used for many other kinds of problems. Here is an example.

Problem 3.1

A number of lines are given in the plane. They subdivide the plane into regions. Prove that you can colour the regions using two colours so that adjacent regions always have different colours.

Here we call two regions adjacent if they have a common border. A common vertex is not enough. See Figure 3.2.

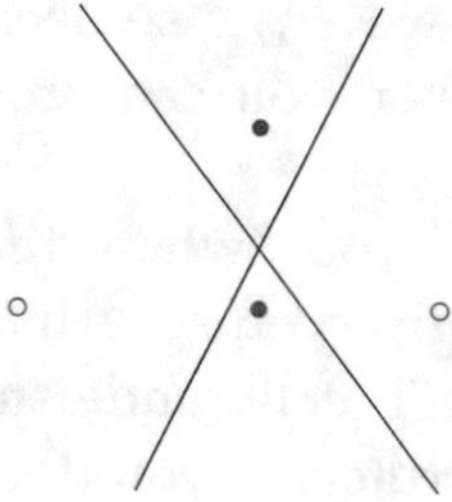

Figure 3.2 Example of a proper colouring

Investigation

Let us call a colouring *proper* if adjacent regions always have different colours. Also, let us call a colouring by two colours a *2-colouring*.

▷ Draw a few **examples** in order to check that the claim has a chance of being true, and to **get a feel for the problem**. Try to find a reason why it always works.

▷ *An idea:* Look at an intersection point of the lines. No matter how many lines run through it, the number of regions having the point as a corner will be even. Therefore we can alternate colours and get a proper 2-colouring for these regions. We can do this at each intersection point. However, we need to make sure that these 'partial' colourings fit together, since most regions will have several corners. How could we prove this? It seems hard to describe in general how the various intersection points fit together.

Since we don't see how to proceed[5] we try something else:

▷ *Second idea: induction with respect to the number of lines.* Denote the number of lines by n. **Let us think about how an inductive proof might work**. What would the inductive step be? Given $n+1$ lines we want to find a proper 2-colouring. In order to use the inductive hypothesis we remove one line and 2-colour the regions formed by the remaining n lines. Then we put the line back in. Now we get new regions, and we have to colour them.

Therefore, the central question in the inductive step will be:

> *Central question:* Suppose n lines and a proper 2-colouring of the regions formed by them are given. Suppose we now add another line l. How can we adjust the original colouring to a proper 2-colouring of the regions formed by all $n+1$ lines? (∗)

We need a *general rule* how to adjust the 2-colouring.

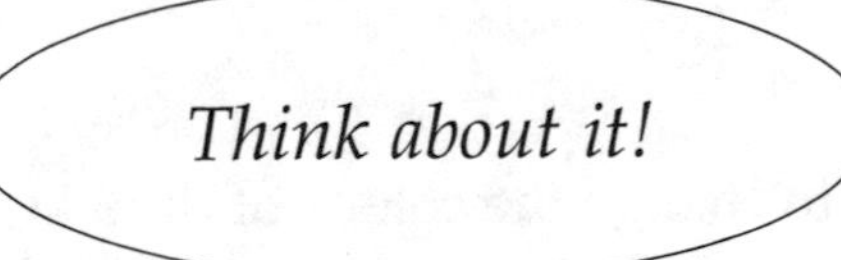

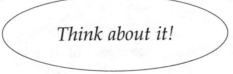

To get an idea, let us investigate the question in the **special case** $n = 2$. We start out with the colouring in Figure 3.2 and add a line l as in Figure 3.3. We can leave the regions below l coloured as before. Above l we get three new regions, and their colours must be chosen as shown. What could be a general rule?

> *First attempt at a general rule:* In any region which is cut by l change the colour on one side of l (always the same side, e.g. above l).

(Since we are using only two colours, it is clear what 'changing' the colour means: use the other colour.)

▷ Is this a good rule? That is, will we always obtain a proper colouring in this way? Figure 3.4 shows that this is not the case. We see that it is not sufficient to change the colour of regions cut by l:

[5] But see Exercise E 3.9 for a solution based on this idea.

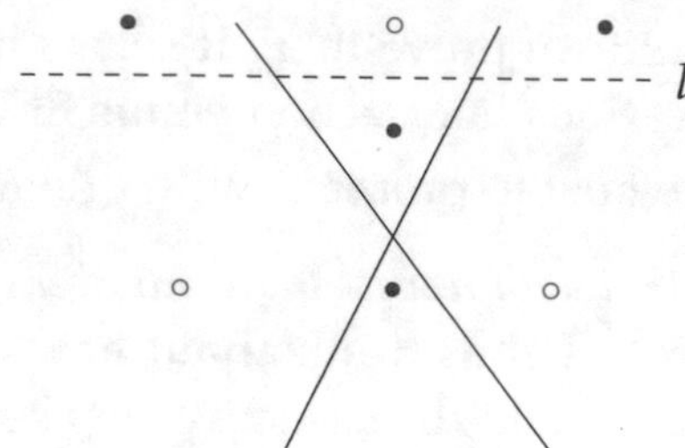

Figure 3.3 New line, I

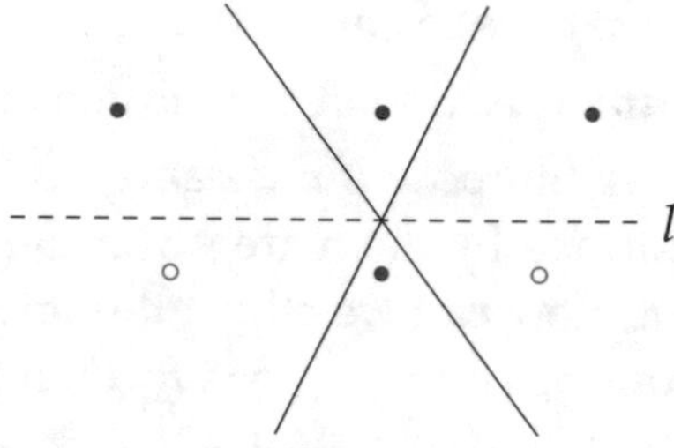

Figure 3.4 New line, II, improper colouring

We also need to change the colour of the top middle region as in Figure 3.5.

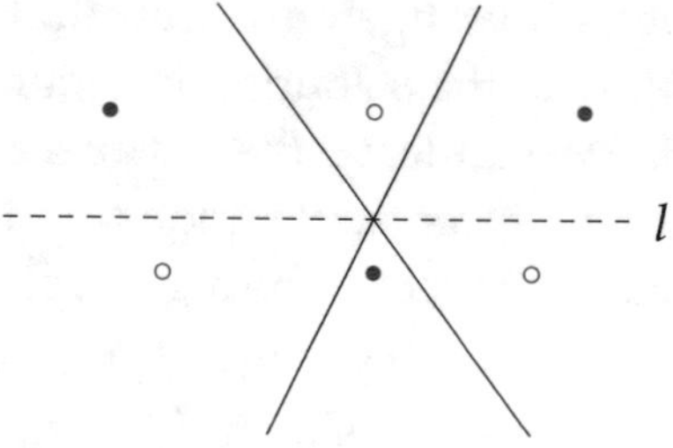

Figure 3.5 New line, II, proper colouring

▷ So our first rule doesn't work in general. We need to modify it. What could be a rule which also works for the line l in the second example?

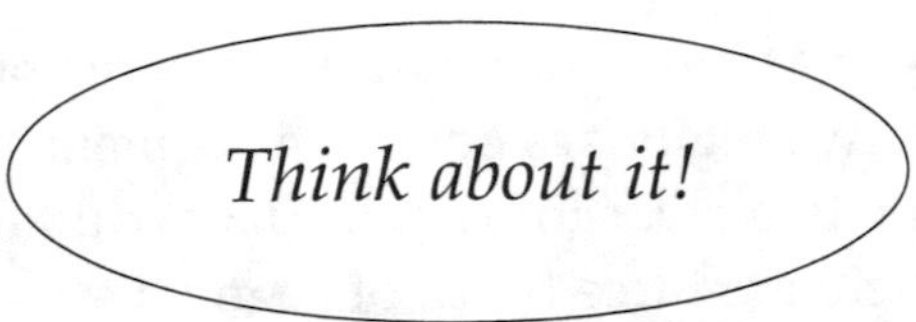

Second attempt at a general rule: Change the colour of every region above l.

Check that this is what we did in the examples, and try to argue that this will *always* lead to a proper colouring! The full argument is written down below.

We now write up the full proof by induction. Pay particular attention to the words **every** and **arbitrary**.

! Solution of Problem 3.1

Let us call a configuration of lines in the plane and the regions formed by them a *map*. Let $A(n)$ be the statement: *Every* map formed by n lines has a proper 2-colouring.

Base case: $A(0)$ is obviously true: If there is no line then there is only one region, and we only need one colour.

Inductive hypothesis: Let $n \geq 0$ be arbitrary. Suppose *every* map formed by n lines has a proper 2-colouring.

Inductive claim: We want to prove $A(n+1)$, that is: *Every* map formed by $n+1$ lines has a proper 2-colouring.

Inductive step: Let an *arbitrary* map M formed by $n+1$ lines be given. We choose one of the lines and call it l. When we remove l we obtain a map with n lines, which we call M'. By the inductive hypothesis M' has a proper 2-colouring. Let us call this colouring C'.

From C' we construct a 2-colouring C of M as follows (this is the answer to question $(*)$ above).

l divides the plane into two half planes. We choose one of them and call it H. Now we start with the colouring C' and change the colours of all the regions contained in H. This results in a 2-colouring of M which we call C.

Claim: The colouring C is proper.

Proof: The lines of M are divided into segments by other lines. Any border between two regions is one such segment. We check for each segment that the two adjacent regions have different colours in the colouring C. There are three types of segments:

1. A segment of the line l. This divides a region of M' into two parts, both of which are regions of M. One of them lies in H. Both regions have the same colour in C', but the region lying in H got its colour changed in C. So the two regions have different colours in C.

2. A segment which is not on l and not in the half plane H. Both adjacent regions have the same colours in C as in C', and since C' is proper, they have different colours in C.

3. A segment which lies in the half plane H. Here both adjacent regions lie in H, so their colours were both changed. Since they had different colours in C' they have different colours in C.

We have shown that the new 2-colouring C is proper. This concludes the proof of the inductive step $A(n) \Rightarrow A(n+1)$. !

Review of Problem 3.1

In the inductive step we showed $A(n) \Rightarrow A(n+1)$. However, the argument did not start with $A(n)$, but with "Let an arbitrary map M formed by $n+1$ lines be given". Only later did we use $A(n)$. Make sure that you understand that this makes sense.

We have introduced several **concepts:** proper colouring, 2-colouring, map. This allowed us to avoid lengthy statements such as 'a colouring which satisfies the conditions of the problem' or 'the regions formed by the lines'. **This not only simplifies the written solution but also helps us in thinking about the problem.**

This was a simple example of one of the central elements of the science of mathematics: finding good concepts.

Similarly, using short notation was also useful: M, l, C etc.

In Chapter 4.5 we will take up colourings again.

3.3 Toolbox

You have learned about mathematical induction as a general proof principle. It is useful for proving statements of the form "For all n …".

It is useful to **plan the induction**, especially if you are not sure whether an inductive proof would work. You ask: What would we have to show for the inductive step?

The basic idea is the same as for recurrence relations: in order to solve the problem of size n (prove statement $A(n)$) we look for a way to reduce it to a smaller problem of the same kind (to the statement $A(n-1)$ or $A(n-2)$ or …).

Introducing suitable concepts is a fundamental task of science. Not only does it simplify writing up arguments, but also it helps you to think about a problem.

Exercises

E 3.1 Prove that you can put infinitely many pins into a suitcase :-). 1

E 3.2 Solve the recurrence relation $a_n = a_1 + \cdots + a_{n-1}$ with initial condition $a_1 = 1$ by first calculating a few values of a_n, stating a conjecture and then proving the conjecture by induction. 1-2

E 3.3 Prove that any triangulation of an n-gon (cf. Problem 2.4) has precisely $n-2$ triangles and $n-3$ diagonals. 2

E 3.4 Consider a square board divided into $2^n \times 2^n$ congruent little squares. Suppose that one of the little squares is colored black, all others are white. Prove that it is possible to tile the white area by L-shaped tiles. Here a tile covers precisely three little squares as in Figure 3.6. 2

E 3.5 PASCAL's triangle is a triangular array of numbers. The first six lines are shown in Figure 3.7. You put ones at the left- and right-hand edges, and then each other entry is the sum of the two entries diagonally above it. Calculate the sums of the numbers along the dotted diagonals, find a pattern, state a general rule and prove it by induction. 2

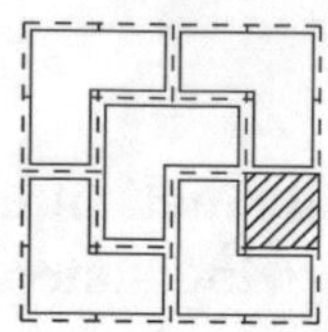

Figure 3.6 Example for Exercise E 3.4 for $n = 2$

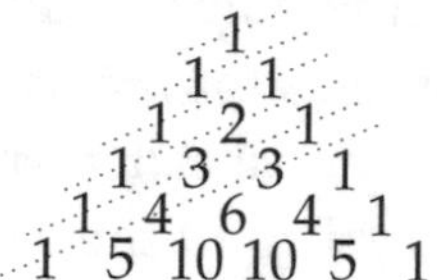

Figure 3.7 PASCAL's triangle, see Exercise E 3.5

2 E 3.6 Analyse the following statement and 'proof'.

Let a be a positive real number and $n \in \mathbb{N} \cup \{0\}$. Then $a^n = 1$.

'Proof': We use induction. For $n = 0$ it is known that $a^0 = 1$. This is the base case.

We now assume that $a^n = a^{n-1} = 1$ and will show that this implies $a^{n+1} = 1$. For this purpose we write

$$a^{n+1} = a^n \cdot a = \frac{a^n \cdot a^n}{a^{n-1}} \overset{(*)}{=} \frac{1 \cdot 1}{1} = 1,$$

where $(*)$ follows from the inductive hypothesis.

By induction the claim follows for all natural numbers n.

3 E 3.7 Let $n \in \mathbb{N}$. Consider subsets $A \subset \{1, \ldots, n\}$ which do not contain any pair of consecutive numbers. Let p_A be the product of the elements of A. We also let $p_\emptyset = 1$. Prove that the sum of all the $(p_A)^2$ over all A equals $(n+1)!$. Table 3.1 shows the sets A and the values of p_A and $(p_A)^2$ in the case $n = 4$. The sum of all $(p_A)^2$ is $5! = 120$.

3 E 3.8 The 'Tower of Hanoi' is a puzzle. You have three vertical rods A, B, C and several disks of different sizes which have holes so you can slide them onto the rods. In the beginning all disks are on rod A,

A	$\emptyset$	$\{1\}$	$\{2\}$	$\{3\}$	$\{4\}$	$\{1,3\}$	$\{1,4\}$	$\{2,4\}$
p_A	1	1	2	3	4	3	4	8
$(p_A)^2$	1	1	4	9	16	9	16	64

Table 3.1 Example for Exercise E 3.7

ordered by size, the largest one at the bottom. Your goal is to move all disks to rod C. In one move you can take a single disk from one rod and put it on another – but only if the disk is smaller than all the disks which are already present on the new rod.

Find a formula for the smallest number of moves which is needed to accomplish this when you have n disks. Prove the formula by induction.

E 3.9 Give another solution for Problem 3.1 by working out the following idea: We choose any region, call it A, and colour it red. Then we colour the other regions as follows: Let B be any region. Choose a path from a point in A to a point in B which does not meet any vertices. Count how often the path intersects one of the given lines (boundaries of regions). If this number is even then B will be coloured red, otherwise blue.

Remark: In this way you can also prove a generalisation of the statement of Problem 3.1: If every vertex of a plane graph has even degree then the faces of the graph can be coloured properly with two colours. (The notions graph, degree and face will be introduced in Chapter 4.)

4 Graphs

The graphs which you will encounter in this chapter are very simple objects at first glance – so simple that most people wouldn't associate them with mathematics. They bear no relation to formulas or equations, nor to geometry. But thinking about them leads to a lot of interesting mathematics, and you will discover some of that mathematics in this chapter. You will use mathematical induction in a new context and learn some new techniques for problem-solving, like counting in two ways and even/odd arguments. You will encounter a proof of impossibility – a fascinating species of mathematical proposition, about the limits of what can be done. Finally, with EULER's formula you will get a first glimpse of the intriguing mathematical area of topology.

4.1 The EULER formula for plane graphs

To understand what a graph is it is best to look at some examples, see Figure 4.1. For the moment we will only look at plane graphs; these are drawn in such a way that the edges don't intersect.[1] Here is a formal definition.

Definition A **plane graph** G consists of:

1. a finite set of points in the plane, the **vertices** of G;
2. finitely many lines joining vertices, the **edges** of G. The edges are not allowed to intersect (except at their endpoints, the vertices).[2]

[1] Be careful: The graphs we are talking about here have no relation to the graphs of functions, that you study in calculus. Therefore they are sometimes called combinatorial graphs.

[2] In most examples the edges are straight lines. However, this is not required in the definition.

D. Grieser, *Exploring Mathematics*, Springer Undergraduate Mathematics Series, https://doi.org/10.1007/978-3-319-90321-7_4

Figure 4.1 Five examples of plane graphs

We assume that there is at least one vertex, but there could be no edges. A plane graph is called **connected** if you can travel from any vertex to any other vertex along edges. A **path** is a sequence of edges where each edge has a common vertex with the previous edge, so that the edges can be traversed one after another. The graphs G_1 to G_4 in Figure 4.1 are connected, G_5 is not.

A plane graph G divides the plane into regions, called the **faces** of G. The 'exterior' of G is also counted as a face. For example, G_1, G_3 and G_4 have two faces each, G_2 has three faces and G_5 has one face.

In the next problem you will encounter EULER's formula, one of the fundamental formulas of mathematics. Its proof is a nice exercise in mathematical induction.

Problem 4.1

Let G be a connected plane graph. Denote by v, e, f the numbers of vertices, edges and faces of G, respectively. Prove that

$$\boxed{v - e + f = 2} \quad \text{(EULER's \textit{formula})} \tag{4.1}$$

Investigation

▷ Look at a few **examples** and check the formula. Also check that the formula is false for disconnected plane graphs. In this way you **get a feel for the formula** and also for the great variety of graphs. Graphs with many vertices can be quite complex. How can we find a proof for the formula in spite of this variety?

▷ *Plan:* We want to prove this inductively. What would the inductive step be? It could run as follows: In order to prove the formula for a graph G we construct from G a smaller graph G', use the inductive hypothesis for G' and then conclude that EULER's formula holds for G.

How can we make a graph smaller? We could remove a vertex or an edge. This would correspond to mathematical induction on v or e, respectively.

▷ *First attempt: Induction on v.* Let us try to do the inductive step.

Inductive hypothesis: Let $v \geq 1$ be arbitrary. Assume that EULER's formula holds for all connected plane graphs having *fewer* than v vertices.

Inductive step: Let G be an arbitrary connected plane graph, having v vertices, e edges and f faces. We want to prove that $v - e + f = 2$.

In order to use the inductive hypothesis we remove a vertex from G; then we also need to remove all edges attached to this vertex. **How do v, e, f change?**

Denote the removed vertex by V and the new graph by G'. Let v', e', f' be the number of edges, vertices, faces of G'. Clearly $v' = v - 1$, but how are e' and f' related to e and f? **To get an idea what happens, we look at some examples,** see Figure 4.2. **However, our arguments must be generalizable to arbitrary connected plane graphs.**

By the inductive hypothesis we have $v' - e' + f' = 2$. (Or not? See below!) We want to conclude $v - e + f = 2$.

▷ The first example in Figure 4.2 is easy: We remove one vertex and one edge, so the difference $v - e$ does not change: $v - e = v' - e'$; also, $f = f'$, so $v - e + f = v' - e' + f' = 2$.

The second example shows that the number of faces can change also. Things get complicated. How can we control the simultaneous change of v, e and f?

Idea: Remove the edges attached to V one by one, and check how f changes in each step: It seems clear that the number of faces decreases by one every time we remove an edge (so $-e + f$ and hence $v - e + f$ remains constant), except when removing the last edge. When removing the last edge we are in the situation of the first example.

▷ This looks promising. But the third example shows that we must be careful: By removing an edge the graph may fall apart, i.e. become

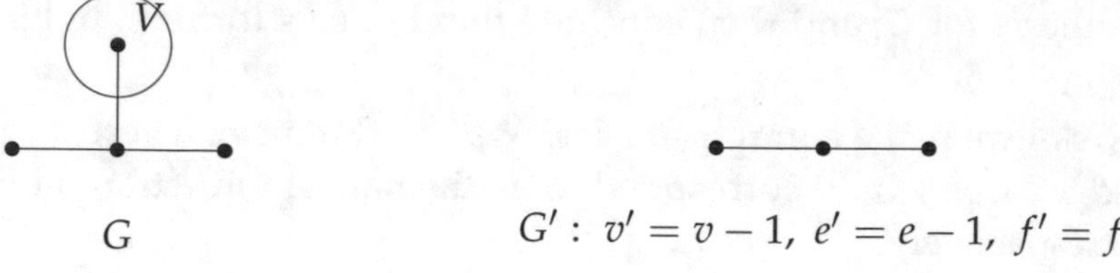

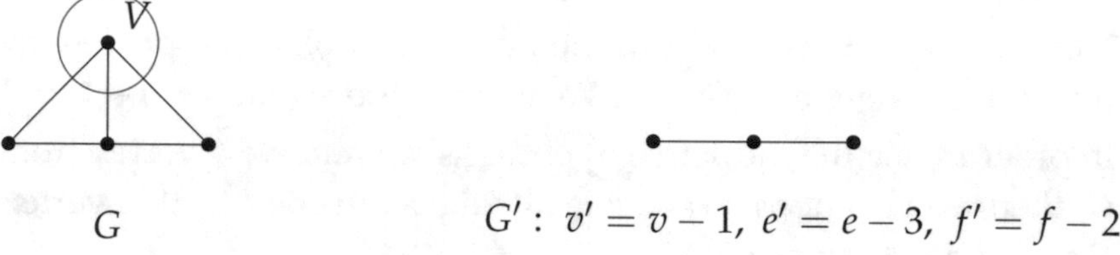

Figure 4.2 Removing a vertex

disconnected.

▷ *Summary of first attempt:* It is not easy to track what happens when removing a vertex. Along the way it is reasonable to remove edges one by one.

This suggest another *idea:* We could start out by removing edges, not vertices. That is, we could do an induction on e, not on v. Try it!

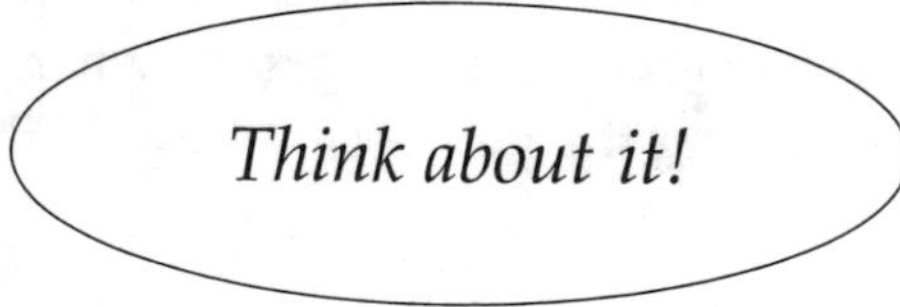

▷ *Second attempt: Induction on e.* Again let us do the inductive step first.

Inductive hypothesis: Let $e \geq 1$ be arbitrary. Assume EULER's formula holds for all connected plane graphs with fewer than e edges.

Inductive step: Let G be an arbitrary graph with e edges, v vertices and f faces.

Remove one of the edges, call it E and the remaining graph G'. Let v', e', f' be the numbers of vertices, edges and faces of G'.

▷ Clearly, $v' = v$ and $e' = e - 1$ since we remove only one edge. (We don't remove any vertices, even if they get separated from the remaining graph by removing E.) This is simpler than in the first attempt, where all three quantities v, e, f could change.

▷ But it is still possible that the graph falls apart when removing E. So we must take this seriously. We consider two cases:

Case 1: G' is connected.

Case 2: G' is not connected.

We make a sketch, see Figures 4.3 and 4.4. The dashed lines represent any path in G or G', the dots more vertices or edges. (The drawn edges on the left and right are examples, they need not be there.)

▷ Let us consider case 1: How does f change? By removing E we amalgamate two faces into one, so we have $f' = f - 1$. Using $k' = k - 1$ we get $e - k + f = e' - k' + f' = 2$ as required.

▷ In case 2 G' is not connected, so we cannot use the inductive

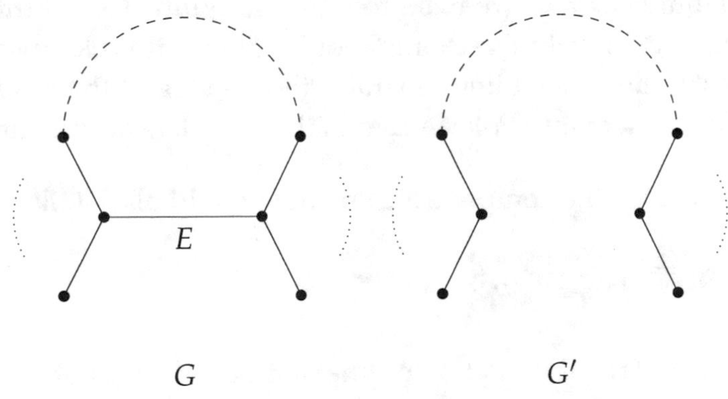

Figure 4.3 Case 1: G' connected

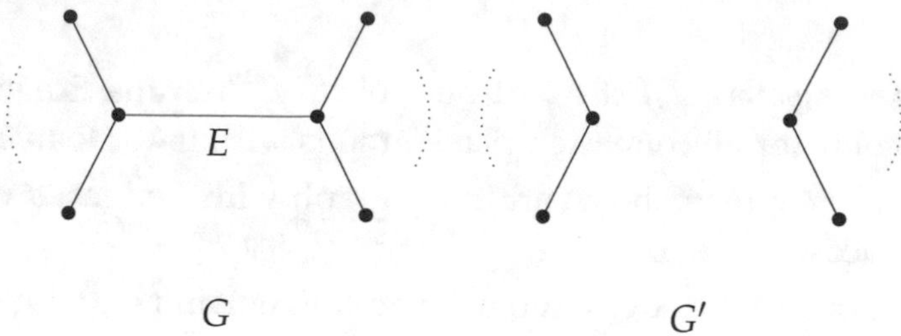

Figure 4.4 Case 2: G' not connected

hypothesis. How can we proceed? One idea is to pull together the two parts and make one vertex out of the two endpoints of E, as in Figure 4.5. Then we obtain a connected plane graph G''. Now we

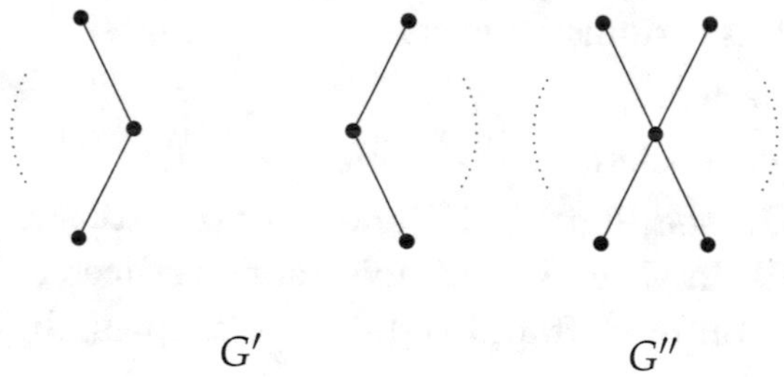

Figure 4.5 Case 2: Gluing the two (former) endpoints of E into one vertex

can use the inductive hypothesis on G''. See below for details. Try it yourself first!

▷ In case 2 the same face borders on E from both sides, so $f' = f$. This reminds us that we need to argue carefully. Could this also happen in case 1? In the example of Figure 4.3 it is clearly not so. Why not? The dashed line separates the upper and the lower side of E. So this fact must play a role in the complete argument.

We now write up the complete argument and add all details.

! Solution of Problem 4.1

We use mathematical induction on the number of edges of G.
Base case: $e = 0$. A connected graph without edges has one vertex and one face, so $v = 1, e = 0, f = 1$, hence $v - e + f = 2$.

Inductive hypothesis: Let $e \geq 1$ be arbitrary. Suppose EULER's formula holds for all connected plane graphs with fewer than e edges.
Inductive step: Let G be an arbitrary connected plane graph with $e \geq 1$ edges, v vertices and f faces.
Remove an edge E from G, and denote the resulting plane graph by G'. Then G' has $e' = e - 1$ edges and $v' = v$ vertices. Let f' be the number of faces of G'.
We distinguish two cases.

Case 1: G' is connected. We first show that E borders two *different* faces of G. Let u, v be the vertices connected by E. If $u = v$ then E will encircle a face, and the claim is obvious. Now assume $u \neq v$. Since G' is connected there is a path in G' from u to v. When combining this path with E we obtain a closed path[3] P. One face bordered by E must lie in the region enclosed by P and one face must lie outside. Therefore they must be different faces.
In G' the two faces bordered by E become one face, all other faces remain unchanged. Therefore $f' = f - 1$. Because $e' < e$ we can use the inductive hypothesis for G' and obtain

$$2 = v' - e' + f' = v - (e - 1) + (f - 1) = v - e + f.$$

Case 2: G' is not connected. Then both sides of the edge E border the same face. Therefore

$$v' = v \quad e' = e - 1, \quad f' = f\,.$$

We form a new connected graph G'' by gluing the two endpoints of E into one new vertex. This does not change the number of edges or faces, but one vertex disappears, therefore, with obvious notation,

$$v'' = v' - 1 = v - 1, \quad e'' = e' = e - 1, \quad f'' = f' = f\,.$$

Because $e'' < e$ we can use the inductive hypothesis for G'' and obtain

$$2 = v'' - e'' + f'' = (v - 1) - (e - 1) + f = v - e + f.$$

In each case we have proven $v - e + f = 2$. This concludes the inductive step.

[3] A **closed path** is a path whose first and last vertex are the same.

Review of Problem 4.1

We first tried induction on v. For the inductive step we removed one vertex. Then we had to remove all edges attached to this vertex. If there were several such edges we had to analyse the simultaneous changes of v, e, f. This suggested that we try an induction on e instead. This was simpler since v does not change when we remove an edge.

Stay flexible! Be prepared to change your route if you get a better idea!

Remark

Here is an alternative solution for the second case:

Since G was connected and we only removed one edge, G' consists of two connected parts, G_1' and G_2' (the left- and right-hand pieces of G' in Figure 4.4). Also, both sides of E border the same face (the 'exterior' face in the figure). Let v_1', e_1', f_1' be the numbers of vertices, edges and faces of G_1' and v_2', e_2', f_2' those of G_2'. Then

$$v_1' + v_2' = v, \quad e_1' + e_2' = e - 1.$$

In addition, we have $f_1' + f_2' = f + 1$ since the exterior face is counted both for G_1' and for G_2'.

Because $e_1' < e$, $e_2' < e$ we can use the inductive hypothesis for G_1' and G_2'. We add and obtain:

$$\begin{array}{rcccccccc} & v_1' & - & e_1' & + & f_1' & = & 2 \\ + & v_2' & - & e_2' & + & f_2' & = & 2 \\ \hline & v & - & (e-1) & + & (f+1) & = & 4. \end{array}$$

Subtracting 2 yields $v - e + f = 2$.

4.2 Counting in two ways for graphs

Counting in two ways is a simple powerful idea that you can use in many contexts to derive interesting formulas.

? Problem 4.2

Let G be a plane graph. Put a mark (a 'custom house') at both sides of each edge, see Figure 4.6. Count the number of custom houses in two ways.

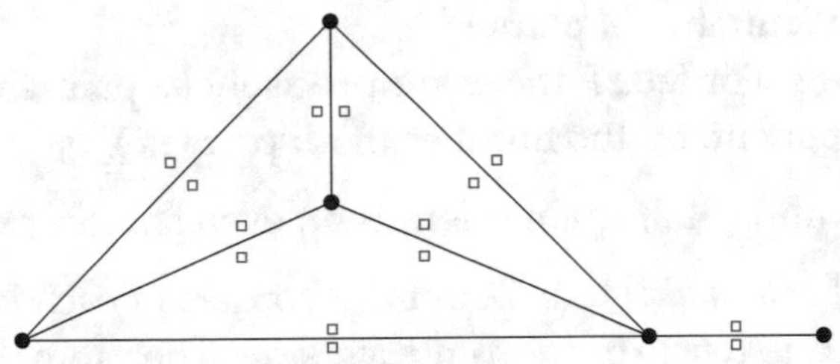

Figure 4.6 How many custom houses?

! Solution

First count: There are two houses at each edge. Therefore there are $2e$ houses.

Second count: Count the houses in each face ('country') and add. To write a formula we use the notation

$$b_F = \text{ the number of borders of the face (country) } F.$$

Then the total number of houses is $\sum_{F \text{ face of } G} b_F$.

Note that the single edge on the right is counted twice as a border of the exterior face. So the exterior face has 5 borders, the other faces have 3 borders each. !

What follows? Since we counted one thing in two ways, the results must be the same, and we get the *edge-face formula*

$$\boxed{2e = \sum_{F \text{ face of } G} b_F} \tag{4.2}$$

In the example: $2 \cdot 7 = 5 + 3 + 3 + 3$. You will find interesting applications of this formula in Problem 4.6 and in Exercise E 4.4.

If you don't like custom houses, say 'one side of an edge' instead, or even better argue like this:

Alternative proof of formula (4.2): We count the pairs (E, F) where E is an edge and F is a face bordered by E. When both sides of E border the same face F we count (E, F) twice.

First count: For each edge E there are two such pairs (or one counted twice). So the number of pairs is $2e$.

Second count: For face F there are precisely b_F pairs that have F as second component. So the number of all pairs is $\sum_F b_F$.

We get formula (4.2) again. Here is an even shorter proof.

Third proof of formula (4.2): Count the borders of each face. You will count each edge twice, once from each side. Therefore we get (4.2).

Since this worked so well we try it again.

? Problem 4.3

Let G be a plane graph. Draw an arrow wherever an edge leaves a vertex, see Figure 4.7. Count the number of arrows in two ways and deduce a formula.

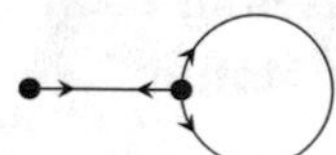

Figure 4.7 How many arrows?

! Solution

First count: Each edge carries two arrows. Therefore there are $2e$ arrows.

Second count: For each vertex V there are as many arrows as there are edges containing V. We let

$$d_V = \text{number of edges containing } V \tag{4.3}$$

(where we count an edge twice if both its endpoints are the vertex V). Then the number of arrows is $\sum_{V \text{ vertex of } G} d_V$. From the two counts we get the *edge-vertex formula*

$$\boxed{2e = \sum_{V \text{ vertex of } G} d_V} \tag{4.4}$$

!

Here it was not essential that the graph was plane. The argument still works if edges intersect. It is also not important how the graph is drawn. This motivates the general definition of a graph, which does not refer to any concrete picture of it.

Definition A **graph** is given by a finite set $\mathcal{V}$ whose elements we call **vertices**, a finite set $\mathcal{E}$ whose elements we call **edges**, and a rule that associates to each $E \in \mathcal{E}$ two vertices $V_1, V_2 \in \mathcal{V}$. It is permitted that $V_1 = V_2$. We call V_1, V_2 the **endpoints** of E and say that E **joins** the vertices V_1, V_2. If $V_1 = V_2$ then we call E a **loop**.

The number d_V defined in (4.3) for a vertex V is called the **degree** of V.

Often we visualise a graph as a diagram in which vertices are given by points in the plane and each edge is represented by a line or curve joining its endpoints but not meeting any other vertices. A plane graph is a representation in which edges don't cross.

Figure 4.8 shows two representations of the same graph: four vertices, with any two vertices joined by a single edge. The left-hand representation is a plane graph, the one on the right is not.

Graphs are a useful tool for visualising complex structures. You will find examples in Problems 4.4 and 10.4.

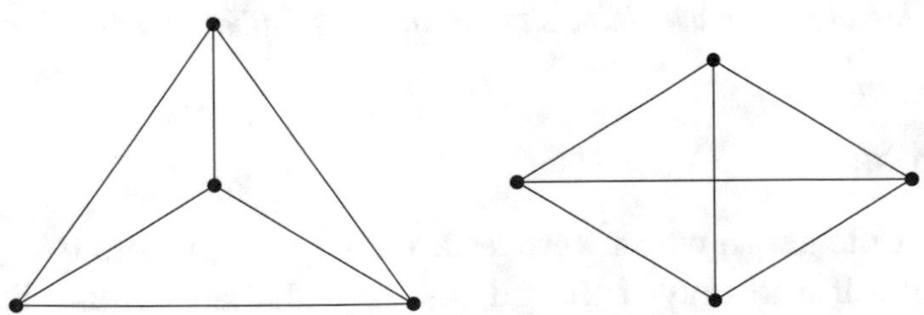

Figure 4.8 Two representations of the same graph

We summarize:

Graph formulas

For a graph with e edges we have

$$2e = \sum_{V \text{ vertex of } G} d_V .$$

For a plane graph with e edges we have

$$2e = \sum_{F \text{ face of } G} b_F .$$

For a plane connected graph with v vertices, e edges and f faces we have EULER's formula

$$v - e + f = 2 .$$

You should know these formulas well. They can often be used for problem-solving.

4.3 Handshakes and graphs

The following problem shows how to model a problem using graphs and how to use formula (4.4).

? Problem 4.4

At a party some of the guests shake hands. Show that at any point in time the number of guests who have shaken an odd number of hands is even.

! Solution

Consider the graph whose vertices are the guests, where two vertices are joined if and only if the guests have shaken hands. For each guest V the degree d_V is the number of handshakes she has made. By formula (4.4) the sum of all d_V is $2e$, an even number. Therefore the number of vertices (guests) V for which d_V is odd must be even. !

4.4 Five points in the plane, all joined by edges

How can we decide whether a given graph can be represented as a plane graph? Let us start with a little warm-up problem.

Problem 4.5

In Figure 4.9, can you connect boxes A with A, B with B and C with C so that the connecting lines don't intersect?

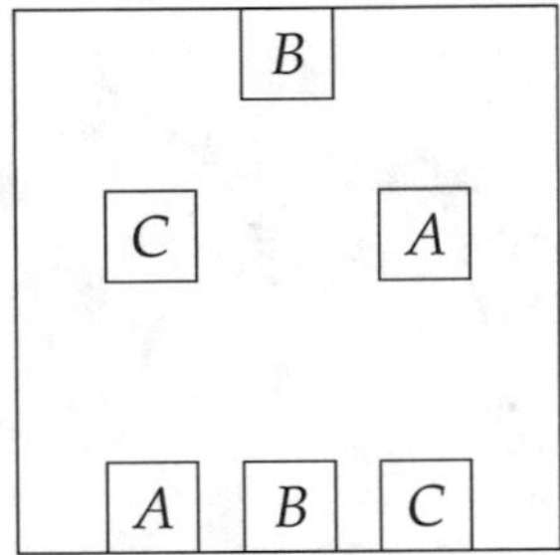

Figure 4.9 Looking for connecting lines

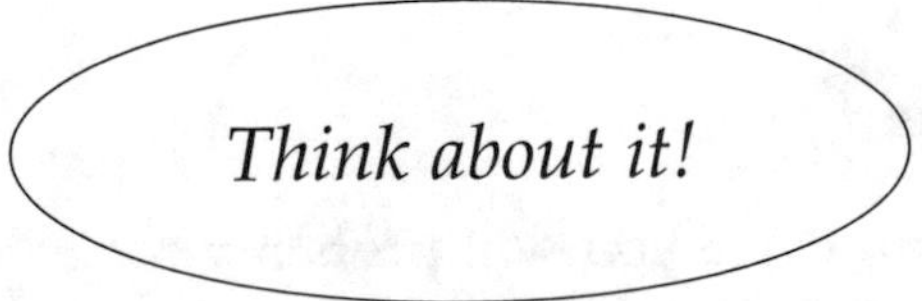

Investigation and solution

At first it looks impossible. Let us try to **simplify**. But consider this: first interchange the two upper boxes A, C, then connect the upper boxes with the lower boxes by vertical lines and finally pull A and C to their original positions (pulling the connecting lines along like rubber bands). This shows that it is possible!

Review

The problem shows that we must be careful not to jump to conclusions of the kind "This is clearly impossible!".

Let us try a more difficult problem.

Problem 4.6

Can you draw 5 points in the plane with a line joining each pair of them, so that the joining lines don't intersect?

Put differently: Does the graph G given by 5 vertices, with any two vertices joined by a single edge, have a plane representation? Recall that with 4 vertices this is possible, see the left-hand picture in Figure 4.8.

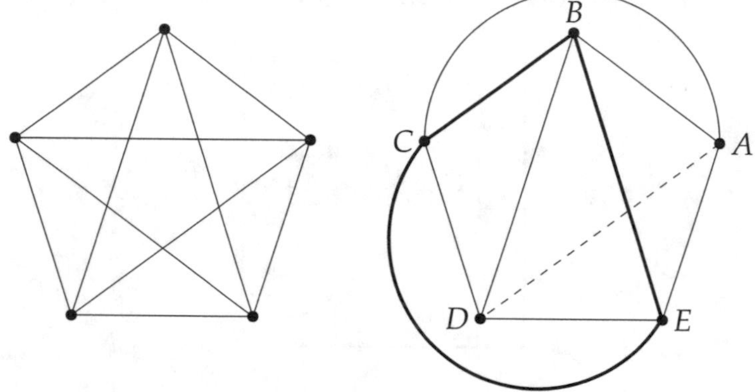

Figure 4.10 A non-plane representation of G and the failed attempt at a plane representation

Investigation

▷ Try it! – After a while you will probably conjecture that it is not possible. For example, in Figure 4.10 all lines except the one from A to D are drawn, and it is clear that we cannot draw the missing line without intersections. This is because A lies outside the closed path $BCEB$ while D lies inside. So we formulate the **conjecture:** It is impossible.

▷ But watch out: The picture does not prove the impossibility. Perhaps we chose the previous lines badly, and if we had started out differently then it would have worked. Problem 4.5 showed us that we need to be careful. **We need a general argument,** one that works no matter how we try to draw the lines.

▷ What do we want to prove? An impossibility. How do you prove that something is impossible? You use a **proof by contradiction:** You suppose that it is possible, and from this you derive a contradiction or a conclusion which is obviously false.

▷ So let us assume that we have a plane representation of G. What can we say about it? Let us **collect information about** G. For example: What are v, e, f? How many borders do the faces have?

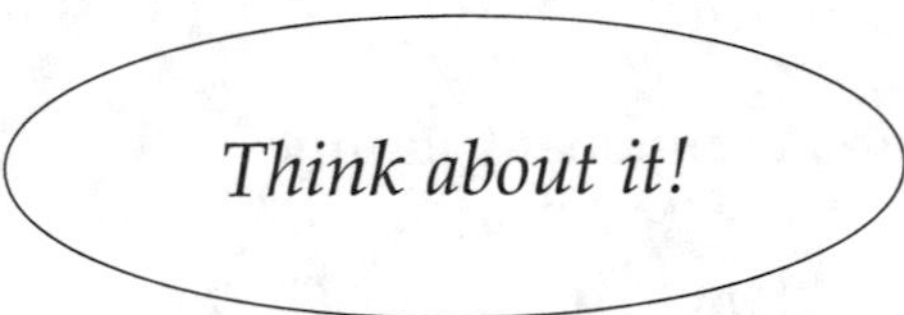

▷ We know that the number of vertices is $v = 5$. Also, we have $e = 10$ since there are 10 ways to choose two out of five vertices (see the left-hand part of Figure 4.10).

▷ What next? Recall our graph formulas. EULER's formula can be used since G is a plane connected graph, and it tells us that $e - k + f = 2$. From $e = 5$, $k = 10$ we get $f = 7$.

▷ What else could we use? For example the vertex-edge formula (4.4). What are the vertex degrees? Each vertex is contained in 4 edges – one to every other vertex. So $d_V = 4$ for all V. The formula gives $2 \times 10 = 4 + 4 + 4 + 4 + 4$, that is $20 = 20$. That's no news.

▷ So let us try to use the edge-face formula (4.2). The left-hand side is $2e = 20$. Can we say anything about the right-hand side, i.e. about the numbers b_F? What can b_F be? Could we have $b_F = 2$ for example, i.e. could there be a face with two borders? – No. This would mean that the two borders have the same endpoints, see Figure 4.11. But there is only *one* edge joining any two vertices. Similarly $b_F = 1$ is impossible since it would mean that the face F is surrounded by a loop; but there are no loops in G.

Figure 4.11 Faces with one or two borders

▷ Therefore we have $b_F \geq 3$ for all faces F. Putting everything together we get

$$20 = 2e = \sum_{F \text{ face of } G} b_F \geq \sum_{F \text{ face of } G} 3 = 7 \cdot 3 = 21\,, \qquad (4.5)$$

where we used that G has $f = 7$ faces. So we derived the false conclusion $20 \geq 21$ from the assumption that G has a plane representation. Therefore this assumption must have been wrong.

Let us write up the argument cleanly.

Solution of Problem 4.6

This is impossible, i.e. there is no plane graph with 5 vertices where any two vertices are joined by an edge.

Proof.
We prove this by contradiction. Suppose there was such a plane graph G. Let v, e, f be the numbers of vertices, edges and faces of G. By assumption we have $v = 5$ and $e = 10$. Because G is plane and connected, EULER's formula $v - e + f = 2$ implies $f = 7$. Because G has no loops and no double edges, every face has at least three borders, see Figure 4.11. The edge-face formula now gives the equations and inequalities (4.5), so $20 \geq 21$. Because this is obviously false, our assumptions that there is such a plane graph G must have been wrong. q. e. d.

Review of Problem 4.6

This was a first example of a *proof of impossibility:*[4] We were able to prove that every one of the infinitely many possibilities of drawing G must have intersecting edges. Our argument was **indirect:** By assuming there were no intersections, we derived a false statement.

4.5 Going further: EULER's formula for polyhedra, topology and the four colour problem

Graphs and polyhedra

EULER's formula for graphs is a generalisation of the following:

[4] In Chapters 7 and 11 you will find more.

Euler's formula for polyhedra: Let v,e,f be the number of vertices, edges and faces of a convex polyhedron in space. Then

$$v - e + f = 2\,.$$

A **convex polyhedron** is a convex three-dimensional body bounded by planar faces.[5] Some examples of convex polyhedra are the cube, the tetrahedron, the octahedron, the dodecahedron, the icosahedron, pyramids ..., but not a cylinder. Check for these examples that the formula holds.

How are convex polyhedra related to plane graphs? Think of a wire model of the polyhedron and position a light source just outside one of the faces. It will cast a shadow on a wall which is parallel to that face and on the other side of the polyhedron. This shadow is a plane graph. Faces of the polyhedron correspond to faces of the graph, with the face where the light source is positioned corresponding to the exterior face of the graph. Therefore, EULER's formula for polyhedra follows from EULER's formula for graphs. The left-hand picture in Figure 4.8 shows the shadow of a tetrahedron (three-sided pyramid). Draw the shadows of the cube and the dodecahedron (twelve pentagons)!

You would get the same plane graph by making the surface of the polyhedron out of rubber, cutting out one face and then pulling everything flat.

By the same argument everything that we derived for plane graphs is also true for graphs that can be drawn on a sphere (surface of a ball) without intersections.

A short excursion into topology

EULER's formula is the starting point for the mathematical area of **topology**. The basic question of topology is: how can you distinguish a sphere (surface of a ball) from a torus (bicycle tire)? More precisely: How can you express mathematically that the torus has a 'hole' that

[5] Convex polyhedra can also be characterized as: 1. The convex hull of finitely many points in space, i.e. the smallest convex set containing these points. 2. The intersection of finitely many half spaces, assuming this is bounded. The equivalence of these conditions is intuitively clear but not easy to prove.

you can put your arm through, while the sphere doesn't – and that this property is preserved when we deform these surfaces continuously?[6]

EULER's formula allows us to find an answer to this question, as follows:

1. We consider (connected) graphs that can be drawn on the sphere or on the torus without intersections. As explained above graphs on a sphere are like plane graphs, for example the formula $v - e + f = 2$ holds. However, for graphs on the torus this is not the case.

2. For connected graphs on the torus there is also a version of EULER's formula. But its right hand side is 0, not 2. Also, the formula only holds for connected graphs on the torus that satisfy an additional condition: Every face must be such that it can be drawn in the plane.[7] For such graphs on the torus we have

   EULER's formula for graphs on the torus: $v - e + f = 0$.

   See Figure 4.12 for an example. This is to be understood as follows: Consider the square of paper whose corners are the four points marked A. Glue together the left and right sides (with the single arrows), to get a cylinder. Then glue together the upper and lower sides (with the double arrows) – stretching the paper as necessary –, to get a torus. The four points marked A will be glued into a single point of the torus, and the same is true for the two Bs and the two Cs. Therefore the graph on the torus has 4 vertices, 8 edges (four containing D, two connecting A and B – not four since the upper and lower one are glued together – and two connecting A and C) and 4 faces, so $v - e + f = 4 - 8 + 4 = 0$. See Exercise E 4.8 for more on this.

3. There are also versions of EULER's formula for other surfaces. For a 'torus with two holes' you get the formula $v - e + f = -2$, for three holes (surface of a pretzel) you get $v - e + f = -4$. In general

[6] Think of turning the sphere into a long sausage, then knotting the sausage. Topologically this is still like a sphere, but what is the essential difference from a torus, which may also be knotted?

[7] For example this is not the case for the graph consisting of a single vertex and no edges, or of two vertices connected by a single edge. When drawing this on a torus the resulting face can not be flattened into the plane.

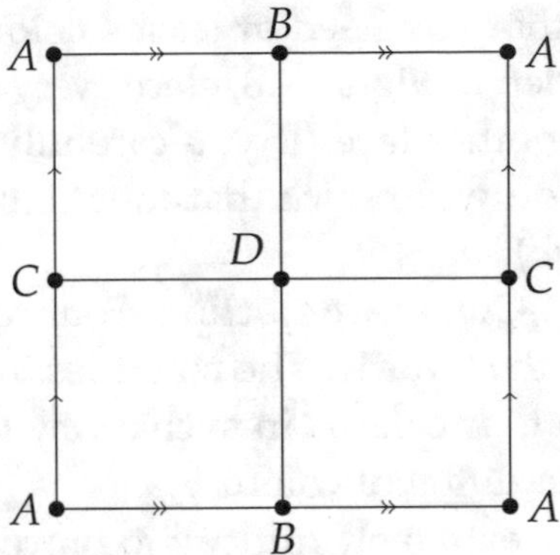

Figure 4.12 A graph on the torus

for g holes you get $v - e + f = 2 - 2g$. The number that you get on the right is called the EULER **characteristic** of the surface.

4. The EULER characteristic answers our question above: It is a well-defined quantity that can be computed for any surface and that 'measures' a difference between sphere and torus.[8] Also, it does not change under deformations of the surface. Therefore, the EULER characteristic is called a **topological invariant.**[9]

See the References section at the end of the book for further reading on topology.

Colourings of plane graphs

? Problem 4.7

Suppose we want to colour the faces of a plane graph in such a way that any two faces which share a border must get different colours. We call such a colouring proper. *What is the smallest number of colours which is enough for a proper colouring of any plane graph?*

---

[8]You may find this unsatisfying: While the EULER characteristic distinguishes sphere and torus it does not intuitively reflect the 'hole' in the torus. But it has the advantage of being easily defined. There is a way to talk about holes mathematically. It is called homology theory, but it takes a lot more effort to define it.

[9]You will learn more about invariants in Chapter 11.

- It is easy to see that you need at least 4 colours, for example for the graph on the left of Figure 4.8, since every one of the four faces borders on every other face. If you carefully try out some more examples then you will notice that four colours always seem to suffice (be enough).
- This leads to the **4-Colour Conjecture:** Four colours suffice for any plane graph. In other words: The countries in any plane map can be coloured with four colours in such a way that countries which share a border get different colours.
- It turned out to be extremely difficult to prove this conjecture. The conjecture was formulated the first time in 1852, but in spite of the efforts by many mathematicians the first proof was only obtained in 1976 (so now it is the **4-Colour Theorem**). Unfortunately part of the proof consists in checking a large number of special cases, too many to do by hand. So a computer was used for this. Even now, no proof is known which does not rely on computer help.[10] Exercises E 4.10 to E 4.14 allow you to retrace some of the steps on the way to the four colour theorem. There you will also see that the analogous problem for the torus is simpler. Here the smallest number of colours is 7.

4.6 Toolbox

While thinking about Problem 4.1 you saw that it is important to **stay flexible,** to change a chosen route if you get a better idea. Problem 4.5 showed that you need to be very careful not to jump to the conclusion that "This is clearly impossible!" – often it is hard or even impossible to have all possibilities in mind. Nevertheless, mathematics allows us sometimes to give rigorous proofs of impossibility, for example using the technique of **proof by contradiction.** You also learned the technique of **counting in two ways** and some of its uses, for example in the proof of impossibility in Problem 4.6.

Finally in Problem 4.4 you saw a simple example of how to use **graphs as representations of complex webs of relationships.**

[10] You find more on the interesting history of this problem in (Aigner, 1987), for example.

Exercises

E 4.1 Can the sum of 111 odd numbers be even? 1

E 4.2 Is it possible that in a group of 57 people each person has 1
exactly three friends in the group? We want to assume that the friendship relation is symmetric, that is: if A is a friend of B then B is a friend of A.

E 4.3 Finish the first attempt at a proof of Euler's formula (induc- 2
tion on v).

E 4.4 Is there a polyhedron with exactly 7 triangular and no other 1-2
faces? (See Section 4.5 for the definition of a polyhedron.)

E 4.5 Imagine a house that has a front door but no back door 2
or other entrance. Can you be sure that there is a room in the house which has an odd number of doors? (Rooms include kitchen, bathrooms, corridors etc.)

E 4.6 Each of three houses needs to be connected to the power 2-3
plant, to the waterworks and to the gasworks. Is it possible to do this so that none of the $3 \times 3 = 9$ connecting lines intersect, that is, none of the wires or pipes cross above or below another?

E 4.7 The surface of a (classical) soccerball is made of pentagonal 2-3
and hexagonal patches which are sewn together at their edges. At each 'vertex' three seams meet. Using this information determine the number of pentagons.
Can you also determine the number of hexagons?[11]

E 4.8 Prove Euler's formula for the torus which is mentioned in 3
Section 4.5.

E 4.9 Find an alternative solution to Problem 4.6. A natural idea is 2
'inside-outside': If it was possible to do it then the connections from A to B, B to C and C to A would form a closed path in the plane which does not intersect itself. Such a path divides the plane into two regions, the inside and the outside, and every curve from an

[11] The same graph structure as for a soccerball appears in the material fullerene, a carbon molecule.

interior point to an exterior point must intersect this closed path.[12] Now consider the possible positions of D and E.

1-2 **E 4.10** Prove that there is no plane graph with five faces each of which borders on every other face. Why does this not immediately imply the 4-Colour Conjecture?

2-3 **E 4.11** Let G be a plane graph all of whose vertices have degree at least 3. Prove that there is a face with at most 5 borders.

3 **E 4.12** Use Exercise E 4.11 to prove the **6-Colour Theorem**: the faces of every plane graph can be coloured properly with 6 colours.

2 **E 4.13** Draw a graph on the torus which has 7 faces, each bordering every other face.

3 **E 4.14** Exercise E 4.13 shows that there is a graph on the torus for which you need 7 colours. Prove the **7-Colour-Theorem for the torus:** the faces of any graph on the torus can be coloured properly with 7 colours.

3 **E 4.15** Consider the left-hand graph in Figure 4.13. Try to draw it in one go, that is, without lifting the pencil and so that you traverse each edge exactly once.[13] There are many ways to do this, but you can only do it if you start at certain vertices. Which vertices are these?

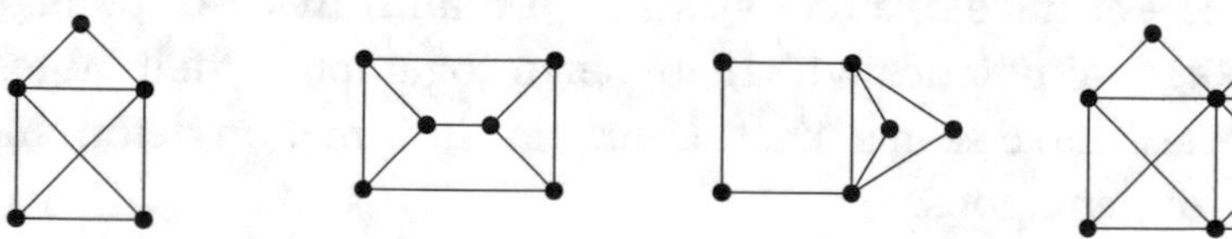

Figure 4.13 The house of Santa Claus and variants

Investigate which of the other figures can be drawn in one go. Can you find a criterion to decide for a given graph whether it can be drawn in one go?

3 **E 4.16** A polyhedron is called **regular** if the same number of edges

[12]This is intuitively clear, but it is not easy to give a rigorous proof. This fact is known as the JORDAN Curve Theorem. You may use it in your argument without proof.

[13]This is a well-known German children's game. While you draw it you say 'Das ist das Haus vom Nikolaus' – 'This is the house of Santa Claus'.

meet at each vertex and every face has the same number of vertices. Prove that cube, octahedron, tetrahedron, dodecahedron and icosahedron are the only regular convex polyhedra.

E 4.17 Label the vertices of a triangle with the numbers 1,2,3. Now 3
introduce some additional vertices in the interior and on the edges, and add edges so that every internal face is a triangle. Then label the interior vertices 1, 2 or 3 however you like, and label each new vertex on an edge of the big triangle with one of the numbers at the end points of this edge.

Prove that there is a (small) triangle all of whose vertices have different labels.

Note: This assertion is known as SPERNER's Lemma. It can be used to prove quite amazing results, for example BROUWER's Fixed Point Theorem.

5 Counting

In how many ways can you arrange 5 objects in a row? How many poker hands have two pairs? Counting is one of the original purposes of mathematics. You can find counting problems in everyday life and in calculating probabilities (how likely is it to have two pairs in a poker hand?). You have already seen some counting problems in previous chapters and learned about the recursion technique. In this chapter we will take a systematic look at counting problems.

Counting is not only a goal in itself, it can also be used to achieve other goals. For example, counting the same objects in two ways can lead to many interesting formulas. Variations of the basic idea of counting in two ways occur in many areas of mathematics, and some of them are explained at the end this chapter.

5.1 The basic principles of counting

Most counting problems can be expressed most clearly in the language of sets and tuples (ordered pairs, triples etc.). If you are not familiar with these notions you should consult Appendix B.

A counting problem can always be stated as counting the elements of a finite set X. We denote

$$|X| = \text{ number of elements of } X\,.$$

This is sometimes called the **cardinality** of X.

The fundamental counting rules

The two fundamental rules of counting are the addition rule and the multiplication rule:

D. Grieser, *Exploring Mathematics*, Springer Undergraduate Mathematics Series, https://doi.org/10.1007/978-3-319-90321-7_5

Fundamental counting rules

1. **Addition rule:** If X is the *disjoint* union of the sets $X_1, X_2, \dots, X_r$,

$$X = X_1 \cup \cdots \cup X_r, \quad X_i \text{ pairwise disjoint,}$$

then $|X| = |X_1| + \cdots + |X_r|$.

2. **Multiplication rule:** If you can specify the elements of X uniquely by first making one decision, with n possible outcomes, and then making another decision, with m possible outcomes (*where m is independent of the outcome of the first decision*) then $|X| = nm$.

   More generally for $s \geq 2$ decisions: If there are n_1 possible outcomes for the first decision, n_2 for the second etc., where *these numbers are independent of all previous decisions*, then $|X| = n_1, \dots, n_s$.

The addition rule is so obvious that is often used without comment.[1] We have already used it in Chapter 2 when we divided the set of things to be counted (e.g. domino tilings, triangulations) into disjoint classes and then counted each class separately.

The multiplication rule follows from the addition rule: Say the possible outcomes of the first decision are $1, \dots, n$. Let X_i be the set of those elements of X where the first decision has outcome i. Then X is the disjoint union of $X_1, \dots, X_n$. Also, $|X_i| = m$ for each i, so $|X| = |X_1| + \cdots + |X_n| = m + \cdots + m = nm$. The rule with $s > 2$ decisions follows similarly by induction on s (exercise!).

An important special case of the multiplication rule is counting the number of elements of the cartesian product of two sets: If A, B are finite sets then

$$|A \times B| = |A| \cdot |B|, \tag{5.1}$$

since in order to specify uniquely an element $(a, b) \in A \times B$, you first choose a in $|A|$ possible ways (first decision) and then for each of these choices you can choose b in $|B|$ possible ways (second decision). Equation (5.1) is also obvious if you arrange the elements of $A \times B$ in a rectangular array as in Appendix B.

[1]If you want a formal proof try mathematical induction.

Examples for the multiplication rule

? Problem 5.1

1. *On a menu there are 3 appetisers, 5 main courses and 2 desserts. How many 3-course meals can you put together?*
2. *How many 2-digit numbers are there?*
3. *How many pairs of numbers (a, b) are there, where $a, b \in \{1, \ldots, 20\}$ and $a \neq b$?*
4. *How many pairs of numbers (a, b) are there, where $a, b \in \{1, \ldots, 20\}$ and $a < b$?*

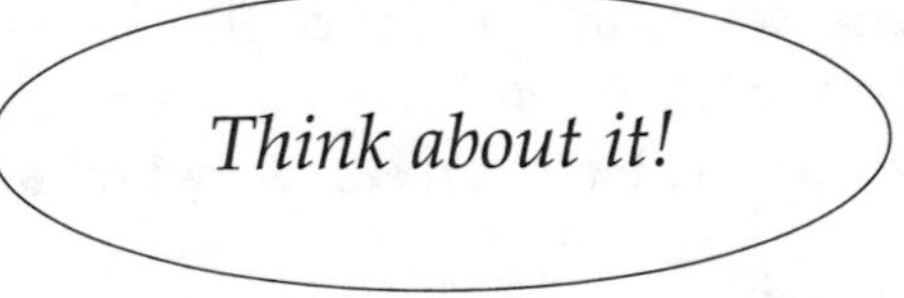

It may be useful for you to write up the possibilities systematically.

! Solution

1. First decision: which appetiser. Second decision: which main course. Third decision: which dessert. Each decision can be made independently of the outcomes of all previous decisions. Therefore there are $3 \cdot 5 \cdot 2 = 30$ different 3-course meals.
2. For choosing the first digit there are the 9 possibilities $1, \ldots, 9$ (first decision), for choosing the second digit there are the 10 possibilities $0, \ldots, 9$ (second decision). So there are $9 \cdot 10 = 90$ numbers.

   Second solution: These are all numbers from 10 to 99, that is, $99 - 10 + 1 = 90$ numbers.[2]

[2]Not $99 - 10$ since you need to include the first and the last one. Compare the shift by one in Problem 1.1.

3. First decision: Choose a. There are 20 possibilities.
   Second decision: Choose b. There are only 19 possibilities since we require $b \neq a$. Therefore, there are $20 \cdot 19 = 380$ pairs.
   Important: *Which* numbers are possible for b depends on the choice of a. However, *how many* numbers are possible for b does not depend on the choice of a. Therefore we can use the multiplication rule, but we are not counting the elements of a product set (as in the special case mentioned above).
   Second solution: There are $20 \cdot 20 = 400$ pairs (a,b) with $a,b \in \{1,\dots,20\}$. From these we subtract the 'forbidden' ones, i.e. those having $a = b$. There are 20 of them, so we get $400 - 20 = 380$ pairs.
4. If we first choose a then the number of possibilities for b depends on what a is: if $a = 1$ then there are 19 possibilities, if $a = 2$ there are 18 etc. Therefore the multiplication rule can not be used. What now? Of course we could calculate $19 + 18 + \cdots + 1 + 0$, but is there a simpler way? Problem 3 was similar and could be solved by the multiplication rule. How can we relate Problem 4 to Problem 3?

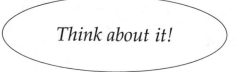

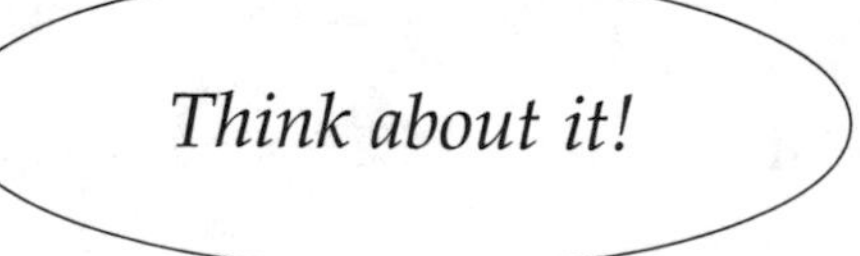

Let

$$X = \{(a,b) : a,b \in \{1,\dots,20\},\ a < b\}$$
$$Y = \{(a,b) : a,b \in \{1,\dots,20\},\ a \neq b\}$$

be the sets to be counted in Problems 4 and 3 respectively. For each pair in X we have two corresponding pairs in Y; for example the pair $(2,5) \in X$ corresponds to the pairs $(2,5),(5,2) \in Y$. Therefore $|Y| = 2|X|$, hence $|X| = |Y|/2 = 380/2 = 190$. !

The idea for solving Problem 5.1.4 was to count Y instead of X, and this amounted to double-counting X. This is often useful:

Multiple-counting principle

If you can't count something, try multiple-counting.

We can make this precise as follows: Suppose we want to count a set X. Then we try to find a set Y having the following properties, for some $m \in \mathbb{N}$:

a) We can count Y.

b) To each element of X we can assign a group of m elements of Y in such a way that the groups don't overlap and Y is the union of all groups.

Then $|X| = |Y|/m$.

Proof: There are $|X|$ groups, whence $|Y| = |X| \cdot m$, therefore $|X| = |Y|/m$.

In the solution of Problem 5.1.4 we assign to $(a, b) \in X$ the elements $(a, b), (b, a) \in Y$. See Figure 5.1 for the analogous problem with $a, b \in \{1, 2, 3\}$.

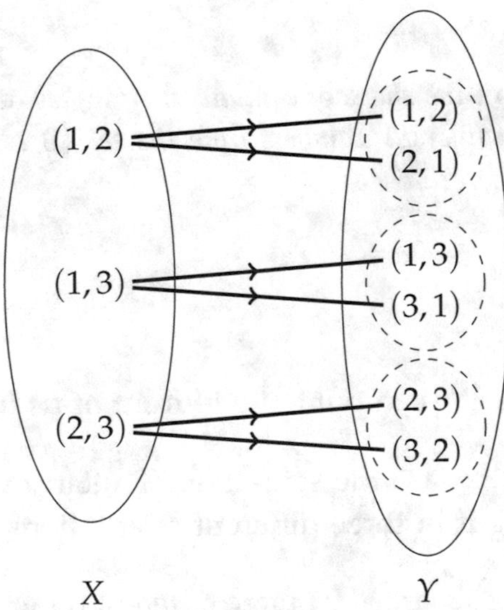

Figure 5.1 Double counting of the pairs (a, b) with $a, b \in \{1, 2, 3\}$ and $a < b$

Additional remarks: The assignment in b) is not a map in the mathematical sense (see Appendix B) since a map assigns a unique image element, not several, to each element of its domain of definition. But if we turn round the arrows in Figure 5.1

then we do obtain a map! Therefore we can formulate condition b) as follows:

There is a map $f : Y \to X$ with the property that for each $x \in X$ the preimage $f^{-1}(\{x\}) := \{y \in Y : f(y) = x\}$ has precisely m elements.

Then the 'groups' of b) are precisely these preimages. Note that the other conditions in b) – the groups don't overlap and their union is all of Y – are satisfied automatically since f is a map. As an exercise you should check this in detail.

How can we write down the map f for Problem 5.1.4? f maps the pair $(a,b) \in Y$ to the pair $(\min\{a,b\}, \max\{a,b\}) \in X$ where $\min\{a,b\}$ is the smaller and $\max\{a,b\}$ is the bigger of the numbers a, b.

Further examples of the multiple-counting rule

? Problem 5.2

A game uses tiles in the shape of equilateral triangles, whose top faces are marked with three different numbers from the set $\{0,\ldots,5\}$. How many different tiles are there?

! Solution

Imagine the tiles. We can put a tile in front of us in different ways: , , are the same tile. Similarly we can turn any other tile, placing it in three different orientations. So if we count fixed triangles marked by three different numbers then we have counted each tile three times.

Therefore we first count how many ways we can mark a triangle with three different numbers: For the left vertex there are 6 possibilities, then for the right vertex there are 5 and then for the top vertex 4 possibilities. Therefore there are $6 \cdot 5 \cdot 4 = 120$ marked triangles. Since we fixed the left, right and top vertex, the three triangles above are counted as different.

By the multiple-counting principle there are $120/3 = 40$ tiles. In this case X is the set of possible tiles and Y is the set of marked triangles. !

? Problem 5.3

The game Triominoes uses tiles as in the previous problem, but the numbers on the tiles need not be distinct. How many different tiles are there?

! Solution

Again we first count how many ways we can mark a fixed triangle. Since the numbers need not be distinct, we have 6 possibilities for each vertex, so by the multiplication rule there are $6 \cdot 6 \cdot 6 = 216$ triangles.
Does every tile correspond to three marked triangles as before?

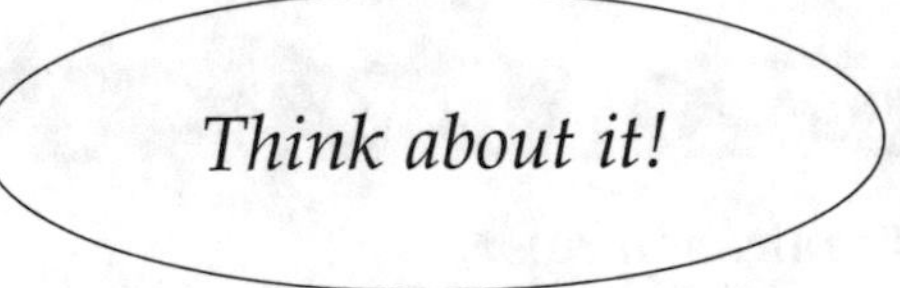

No, since a tile with three equal numbers only corresponds to one marked triangle. There are 6 such tiles. All other tiles correspond to three marked triangles as before. Therefore we apply the multiple-counting principle to those triangles whose three numbers are not all the same. There are $216 - 6 = 210$ such triangles, so $\frac{210}{3} = 70$ corresponding tiles. In addition, there are the 6 tiles with all numbers equal. Altogether we have 76 tiles. !

Review

The multiple-counting principle can only be applied if the size of the multiple-counts (m) is always the same. If this is not the case then one can try to divide the set that is to be counted into subsets (classes) to which the principle is applicable. So here we first used the addition rule and then the multiple-counting rule.

The main counting problems

Using the fundamental counting rules we can solve the most important general counting problems. Many real-life counting problems can be translated into one of these basic types.

Let A be a set with n elements: $|A| = n$. We want to count tuples and subsets of elements of A. We denote the size of the tuple or subset by $k \geq 1$. We always consider the case $k = 2$ first since the principles are most easily understood in this case.

You should work out a special case for each type of counting problem, for example $A = \{1,2,3,4\}$ and $k = 3$. Write out all possibilities and make sure you understand the counting principle.

Number of pairs: $|\{(a,b) : a,b \in A\}| = n^2$

Proof by the multiplication rule:

	number of possibilities for a:	n
and for *each* of them:	number of possibilities for b:	n

This is the special case of equation (5.1) where $A = B$.

Number of k-tuples: $|\{(a_1,\ldots,a_k) : a_1,\ldots,a_k \in A\}| = n^k$

Proof by the multiplication rule:

	number of possibilities for a_1:	n
and for *each* of them:	number of possibilities for a_2:	n
	$\vdots$	
and for *each* of them:	number of possibilities for a_k:	n

Number of pairs with different components:

$$|\{(a,b) : a,b \in A,\ a \neq b\}| = n(n-1)$$

Proof by the multiplication rule:

	number of possibilities for a:	n
and for *each* of them:	number of possibilities for b:	$n-1$

Number of k-tuples with different components:

$$|\{(a_1,\ldots,a_k) : a_1,\ldots,a_k \in A,\ \text{all } a_i \text{ different}\}|$$
$$= n(n-1)\ldots(n-k+1)$$

Proof by the multiplication rule:

	number of possibilities for a_1:	n
and for *each* of them:	number of possibilities for a_2:	$n-1$
	$\vdots$	
and for *each* of them:	number of possibilities for a_k:	$n-k+1$

(This argument is for $k \leq n$, but the formula is also true for $k > n$ since then both sides are equal to zero.)

An important special case is $k = n$:

Number of orderings:

$$|\{\text{Orderings of } n \textit{ different} \text{ objects}\}| = n!$$

Here $n! = 1 \cdot 2 \cdot \dots \cdot n$ (n factorial). Another word for ordering is **permutation.**

Number of 2-element subsets:

$$|\{\text{subsets of } A \text{ with 2 elements}\}| = \frac{n(n-1)}{2}$$

Proof by the multiple-counting principle, compare Problem 5.1.4: Each subset with 2 elements can be written as $\{a,b\}$ where $a < b$. To each such subset we assign the pairs (a,b) and (b,a). Clearly we obtain all pairs with different components, and each such pair arises from a unique subset. Since there are $n(n-1)$ pairs with different components in A we get the formula.
In short: When counting pairs (a,b) with $a \neq b$ instead of subsets then we count each subset twice.

Number of k-element subsets:

$$|\{\text{subsets of } A \text{ with } k \text{ elements}\}| = \frac{n(n-1)\dots(n-k+1)}{k!}$$

Proof by the multiple-counting principle: Each k-element subset of A can be written $\{a_1, a_2, \dots, a_k\}$ with $a_1 < a_2 < \dots < a_k$. To each subset written in this way we assign all k-tuples which can be formed from the a_i in all possible orderings. There are $k!$ such orderings.

In this way we obtain all k-tuples with different components, and each such k-tuple arises from a unique subset. Since there are $n(n-1)\ldots(n-k+1)$ such k-tuples the formula follows by the multiple-countingprinciple where $m = k!$.[3]

Since the expression in the last problem appears frequently, there is a short notation for it.

Definition Let $n, k \in \mathbb{N}_0$, $0 \leq k \leq n$. The **binomial coefficient** $\binom{n}{k}$ (read: n choose k) is defined for $k \geq 1$ by

$$\binom{n}{k} = \frac{n(n-1)\ldots(n-k+1)}{k!}.$$

For $k = 0$ we define $\binom{n}{0} = 1$.

Examples: $\binom{n}{1} = n$, $\binom{n}{2} = \frac{n(n-1)}{2}$, $\binom{n}{3} = \frac{n(n-1)(n-2)}{6}$ etc.[4]

Every set has precisely one 0-element subset (the empty set), so using the last result above we get for all k, n with $0 \leq k \leq n$:

Number of k-element subsets of an n-element set: $\binom{n}{k}$

The numbers $\binom{n}{k}$ appear in PASCAL's triangle, which is introduced in Exercise E 3.5, and in the general binomial theorem. See Exercises E 5.14 and E 5.15.

Examples using the main counting problems

Many counting problems can be reduced to one of the basic types. You only need to recognise it! Here are a few examples.

? Problem 5.4

10 *people meet. Each one shakes hands with every other person exactly once. That's how many handshakes?*

[3]The corresponding map – see the additional remarks before Problem 5.2 – is very easy to write down in this case: if $(b_1, \ldots, b_k)$ is a k-tuple with all b_i different then $f(b_1, \ldots, b_k) = \{b_1, \ldots, b_k\}$.

[4]A good way to remember the lengthy expression in the numerator is to notice that it has precisely k factors, just like the denominator.

! Solution

Let A be the set of 10 people. Each handshake corresponds to a 2-element subset of A. So the number of handshakes is $\binom{10}{2} = \frac{10\cdot 9}{2} = 45$.
!

The same counting problem appears in many disguises, for example: number of connections between n points; number of pairs (a, b) with $a, b \in \{1, \ldots, n\}$ and $a < b$ (note that these pairs correspond precisely to 2-element subsets since any such subset can be written as $\{a, b\}$ with $a < b$; compare Problem 5.1.4).

? Problem 5.5

A group of 10 women and 10 men go dancing. How many ways are there to make up 10 mixed pairs?

! Solution

We number the women $1, \ldots, 10$. The first woman can choose among all 10 men, then the second woman can only choose among 9 men (no matter which man was chosen by the first woman), the next one among 8 etc., the last woman has the choice of only one man. Therefore there are $10 \cdot 9 \cdot \cdots \cdot 1 = 10! = 3{,}628{,}800$ possibilities.

Alternatively you could line up the women. Then each pairing corresponds to one way of lining up the men next to the women. So you need to count the orderings of the 10 men. !

5.2 Counting using bijections

There is another important counting rule. To introduce it we need the concept of bijection. A **bijection** between two sets X and Y is a rule that associates with each element of X an element of Y, such that each element of Y appears precisely once.[5] Figure 5.2 shows an example.

The following statement is so obvious that it may surprise you that it is useful at all.

[5] See Appendix B for more information on bijections.

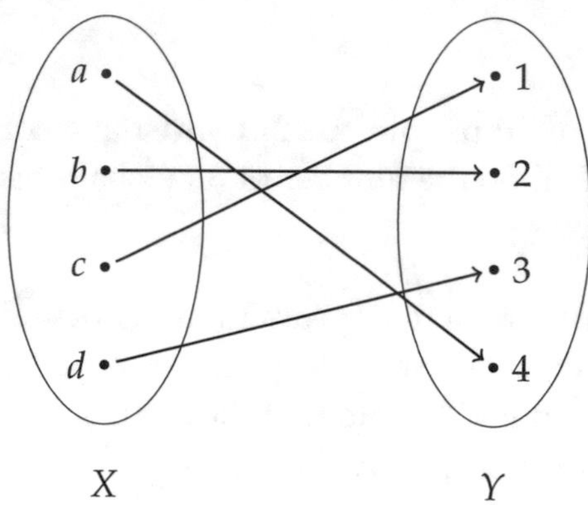

Figure 5.2 A bijection $X \to Y$ where $X = \{a, b, c, d\}$, $Y = \{1, 2, 3, 4\}$

Counting by bijection

If there is a bijection $X \to Y$ then $|X| = |Y|$.

This can be used in two ways:

- For counting: If you want to count X then you can try to find a set Y whose cardinality you know, and a bijection $X \to Y$.

  Even if you already know a formula for $|X|$ such a bijection can lead to a better understanding of the formula.

- For deriving or proving equations: If you have two sets X, Y and expressions for their cardinalities, and if you have a bijection $X \to Y$ then the two expressions must give the same result.

Here are some examples for illustration.

Variations on the number of subsets

We know from Chapter 2.2 that the number of subsets of $\{1, \ldots, n\}$ is 2^n. Recall that we denote the set of subsets of $\{1, \ldots, n\}$ by $\mathcal{P}(\{1, \ldots, n\})$, so

$$|\mathcal{P}(\{1, \ldots, n\})| = 2^n .$$

We conjectured this from a few examples and then proved it using a recurrence relation. Is there a better way to understand the formula?

The number 2^n arises among the basic types of counting problems as the number of n-tuples $(a_1, \ldots, a_n)$, where each a_i takes one of two possible values. Can we relate these n-tuples to the subsets of $\{1, \ldots, n\}$? **Can we find a bijection?**

Let us represent the subsets as follows, for example if $n = 2$:

subset	1	2
$\emptyset$	$-$	$-$
$\{1\}$	$+$	$-$
$\{2\}$	$-$	$+$
$\{1,2\}$	$+$	$+$

For each subset there is one row. In the column marked 1 we write a $+$ whenever the subset contains the element 1, otherwise we write a $-$, and similarly for the column marked 2. If we read the table row by row then we see that each subset corresponds to a pair of elements of $A = \{-, +\}$. For subsets of $\{1, 2, 3\}$ we would get three columns, hence triples. In general we obtain:

Second proof of the equation $|\mathcal{P}(\{1, \ldots, n\})| = 2^n$: The subsets of $\{1, \ldots, n\}$ correspond to the n-tuples $(a_1, \ldots, a_n)$ with $a_i \in \{-, +\}$ for all i, as follows. Each subset $S \subset \{1, \ldots, n\}$ is associated with the n-tuple $(a_1, \ldots, a_n)$ defined by

$$a_i = \begin{cases} + & \text{if } i \in S \\ - & \text{else.} \end{cases}$$

Clearly each n-tuple arises from a unique subset. Therefore the number of subsets equals the number of n-tuples of pluses and minuses, hence is 2^n. q. e. d.

What did we do? In order to count the subsets we associated them with objects that we knew how to count. We constructed a *bijection*

$$\mathcal{P}(\{1, \ldots, n\}) \to \{(a_1, \ldots, a_n) : a_i \in \{-, +\} \text{ for all } i\}$$

(the map was defined above).

By interpreting plus and minus signs as decisions we can reformulate this as follows:

Third proof of the equation $|\mathcal{P}(\{1, \ldots, n\})| = 2^n$: In order to specify a subset S of $\{1, \ldots, n\}$ we first decide whether 1 is in S or not; then we decide whether 2 is in S or not; and so on up to n. These are n independent decisions, and each decision has two possible outcomes. The product rule then implies that there are 2^n subsets. q. e. d.

This gives us a clear understanding *why* the number of subsets is 2^n. The recursive proof given in Problem 2.1 did not have this virtue.

In the following problem we use a bijection for proving a formula, rather than for counting.

? Problem 5.6

Prove the identity

$$\binom{n}{k} = \binom{n}{n-k}$$

(for $0 \leq k \leq n$) using a bijection.

! Solution

The left-hand side counts the k-element subsets of $\{1, \ldots, n\}$, the right-hand side counts the $(n-k)$-element subsets. How can we relate these to each other? Can we find a bijection?

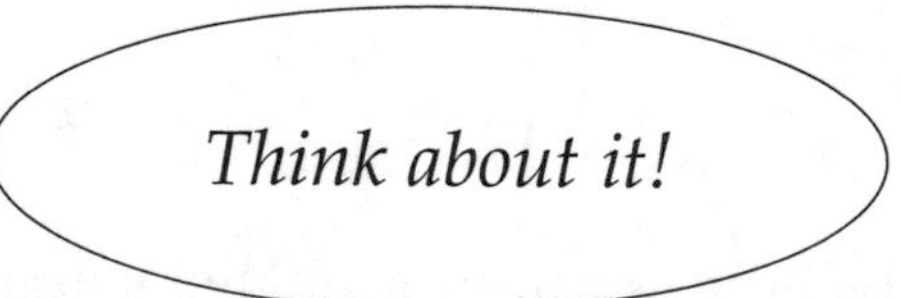

By associating a k-element subset $S \subset \{1, \ldots, n\}$ with its complement $\{1, \ldots, n\} \setminus S$ we obtain a bijection from the set of k-element subsets to the set of $(n-k)$-element subsets. Therefore these two sets have the same cardinality.

More formally: Let

$$X = \{S \subset \{1, \ldots, n\} : |S| = k\}, \; Y = \{S \subset \{1, \ldots, n\} : |S| = n - k\}.$$

Then $F : X \to Y$, $S \mapsto \{1, \ldots, n\} \setminus S$ is a bijection, so $|X| = |Y|$ and hence $\binom{n}{k} = \binom{n}{n-k}$. !

This means that PASCAL's triangle, see Exercise E 3.5, is symmetric.

? Problem 5.7

For $n \geq 1$ let e_n and o_n be the numbers of subsets of $\{1,\ldots,n\}$ which have an even or odd number of elements, respectively. Find formulas for e_n and o_n.

Investigation

▷ Let us look at some examples. We leave out the braces (or "curly brackets") for simplicity. Let us call a subset even/odd if it has an even/odd number of elements.

n	even subsets	odd subsets	e_n	o_n
1	$\emptyset$	1	1	1
2	$\emptyset, 12$	$1, 2$	2	2
3	$\emptyset, 12, 13, 23$	$1, 2, 3, 123$	4	4

▷ We observe that $e_n = o_n$ for $n = 1, 2, 3$. Also, these numbers are powers of 2.

▷ The second of these statements follows from the first: Since every subset must be even or odd and there are a total of 2^n subsets we have $e_n + o_n = 2^n$. So if $e_n = o_n$ then it follows that $e_n = \frac{1}{2}2^n = 2^{n-1}$.

▷ **Conjecture:** We have $e_n = o_n$ for all $n \geq 1$. How could we prove this?

▷ An equality of two numbers suggests that a bijection should exist between the two sets being counted. Denote by E_n, O_n the set of even/odd subsets of $\{1,\ldots,n\}$. We want to **find a bijection**

$$F : E_n \to O_n .$$

What does this mean? We are looking for a rule that associates an odd subset with every even subset. The rule must be such that every odd subset appears precisely once as a result.

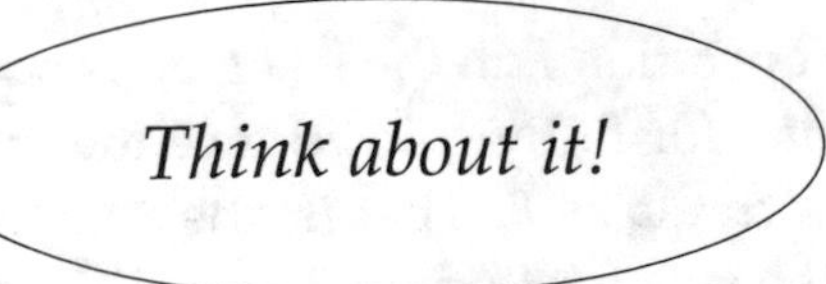

▷ How can you produce one subset from another subset? You know one way to do this from Problem 5.6: taking the complement. That is, associate $S \subset \{1,\ldots,n\}$ with its complement $\{1,\ldots,n\} \setminus S$. Does this rule produce an odd subset from an even one? If n is odd then this is the case, as you can easily check. But if n is even then the complement of an even subset is even again.

So for odd n we have solved our problem.

But what about even n? We need a new idea.

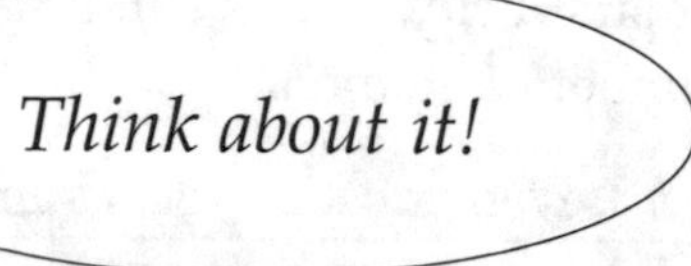

▷ We want to associate odd subsets with even subsets. A **similar problem** would be to associate odd numbers to even numbers. That's easy: add or subtract 1.

▷ Can we do anything similar for sets? We could transform an even subset into an odd one by adding or removing an element. How do we turn this into a **general rule**? When do we add, when do we remove an element? And which element?

Here is a method to do this: If the subset contains the element 1 then take it away, otherwise add it in. That is, our map is defined by

$$F(S) = \begin{cases} S \setminus \{1\}, & \text{if } 1 \in S \\ S \cup \{1\}, & \text{if } 1 \notin S\,. \end{cases} \tag{5.2}$$

Clearly, if S is even then $F(S)$ is odd. We need to check that every odd subset T appears as $F(S)$ for precisely one S. For this we simply invert the operation F: Let T be an odd subset of $\{1,\ldots,n\}$.
If $1 \in T$ then we have $T = F(S)$ for $S = T \setminus \{1\}$;
if $1 \notin T$ then we have $T = F(S)$ for $S = T \cup \{1\}$.
In both cases this is clearly the only possible choice for S.

Therefore we have proved that $F : E_n \to O_n$ is bijective.

▷ Inverting the operation F in the last step is nothing else but finding the inverse map for F. By coincidence the inverse map is defined by the same formula as F. That is, the rule for finding S from T is the same as the rule for finding $T = F(S)$ from S.

▷ This argument works for all n, even and odd.

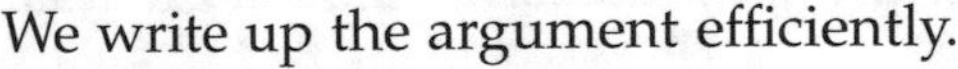

We write up the argument efficiently.

Solution of Problem 5.7

For any subset $S \subset \{1,\ldots,n\}$ define the subset $F(S) \subset \{1,\ldots,n\}$ by formula (5.2). We have $F(F(S)) = S$ for all S, for if $1 \in S$ then $1 \notin F(S)$, so $F(F(S)) = F(S) \cup \{1\} = (S \setminus \{1\}) \cup \{1\} = S$, and similarly if $1 \notin S$ then we get $F(F(S)) = (S \cup \{1\}) \setminus \{1\} = S$.

Therefore $F : \mathcal{P}(\{1,\ldots,n\}) \to \mathcal{P}(\{1,\ldots,n\})$ is its own inverse map. A map having an inverse is bijective. Also, F maps E_n to O_n and O_n to E_n, therefore it defines a bijection $E_n \to O_n$. Therefore we get $e_n = o_n$, and using $e_n + o_n = 2^n$ we get $e_n = o_n = 2^{n-1}$.

Review of Problem 5.7

We looked for a bijection in order to prove $e_n = o_n$. The first attempt (taking complements) only worked for odd n. Then we found another map (rule) that worked for all n.

Lesson: Finding a bijection requires creativity. If the first attempt fails there may still be other ways to find one.

5.3 Counting in two ways

Usually we count in order to determine some number of possibilities. Now we want to use **counting for other purposes.** The starting point is that most problems have many solutions:[6]

> **Counting in two ways**
>
> By counting the same set in two different ways we can often derive interesting formulas.

You have already encountered two examples of this in Chapter 4: the edge-face formula (4.2) and the edge-vertex formula (4.4) for graphs.[7]

[6]Counting in two ways is sometimes called double counting, but this expression is also used for 'overcounting by a factor of two' as in the multiple-counting principle.

[7]Problem 5.6 showed another way of deriving a formula by counting.

? Problem 5.8

Connect n points pairwise by lines. Count the number of lines in two ways and derive an identity.

! Solution

- *First count:* Number the points $1, \ldots, n$.
  - First connect 1 and 2, 1 and 3, …, 1 and n, that's $n-1$ lines,
  - then connect 2 and 3, 2 and 4, …, 2 and n (note that 2 and 1 are already joined), that's $n-2$ lines,
  - then connect 3 to the points $4, \ldots, n$, that's $n-3$ lines,
  - and so on, and finally connect $n-1$ and n.

  Therefore the total number of lines is

$$(n-1) + (n-2) + \cdots + 1\,.$$

- *Second count:* Since we want to draw a line between any two points, the lines correspond to the 2-element subsets of the set of points. There are $\binom{n}{2}$ such subsets. (Compare Problems 5.4 and 5.1.4.)
- Therefore we get

$$(n-1) + (n-2) + \cdots + 1 = \binom{n}{2}\,. \qquad !$$

If we reverse the order on the left and replace n by $n+1$ we obtain

$$1 + 2 + \cdots + n = \binom{n+1}{2} = \frac{(n+1)n}{2}\,.$$

This is a new derivation of the formula for the sum of the first n natural numbers. In Section 1.3 we derived this formula using the Gauss trick.

A similar idea can be used to find a formula for the sum of squares:

? Problem 5.9

Count the number of triples (a, b, c) which satisfy

$$a, b, c \in \{1, \ldots, n\} \quad \textit{and} \quad a < c,\ b < c$$

in two ways and derive an identity.

Investigation and solution

▷ *First count:* c seems to play a special role since it appears in both conditions $a < c$, $b < c$. Therefore it may be useful to sort the triplets by the value of c. How many triplets are there for a fixed value of c?

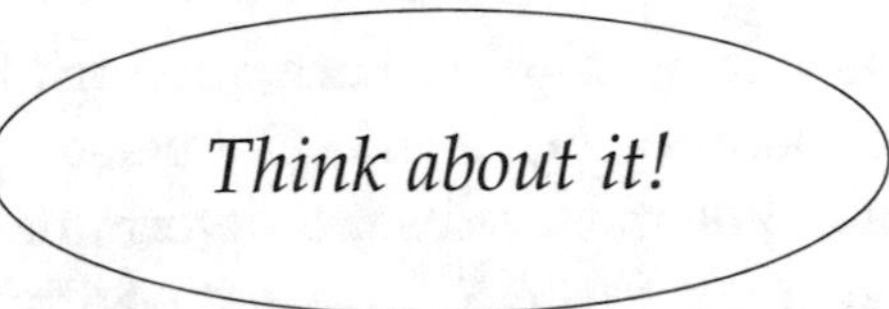

It may help to make a **table**:

c	triples	number
1	none	0
2	(1,1,2)	1
3	(1,1,3), (1,2,3), (2,1,3), (2,2,3)	4

Conjecture?

▷ How many ways of choosing a, b are there for any fixed c? The conditions are $a < c, b < c$, so $a, b \in \{1, \ldots, c-1\}$. Since a, b can be chosen independently, there are $(c-1)^2$ possibilities.

▷ Since any choice of $c \in \{1, \ldots, n\}$ is possible, by the addition rule we get that there are

$$0^2 + 1^2 + 2^2 + \cdots + (n-1)^2$$

triplets.

▷ *Second count:* How else could we count the triplets? Instead of sorting them by the value of c, we could first fix the numbers that appear in the triplet and then count how many triplets can be formed from these numbers. For example, from the numbers 1, 3, 4 we can form the triplets (1, 3, 4) and (3, 1, 4), but from the numbers 2, 5 we can only form the triplet (2, 2, 5).

▷ The examples suggest that we should distinguish whether two or three different numbers appear in a triplet (it is not possible that only one number appears since c is always larger than a and b). Therefore we **divide** the triples **into two classes** and do a **case by case analysis**:

1. The triplets having $a = b$. Each such triplet has two different numbers a and c, so it determines a 2-element subset S of $\{1, \ldots, n\}$. Conversely any 2-element subset S determines a unique triplet: $a = b$ must be the smaller element of S and c must be the larger element. Therefore there are $\binom{n}{2}$ such triplets.
2. The triplets having $a \neq b$. Each such triplet determines a 3-element subset S of $\{1, \ldots, n\}$. Conversely, we saw in the example that a 3-element subset S determines *two* triplets. This works in general: Write $S = \{x, y, z\}$ with $x < y < z$, then S corresponds to the triplets (x, y, z) and (y, x, z). Since there are $\binom{n}{3}$ 3-element subsets S, we get $2\binom{n}{3}$ such triplets.

By the addition rule we get that the total number of triplets is

$$\binom{n}{2} + 2\binom{n}{3}.$$

▷ We obtain the identity

$$1^2 + 2^2 + \cdots + (n-1)^2 = \binom{n}{2} + 2\binom{n}{3}.$$

If we replace n by $n+1$ and calculate a little we get

$$1^2 + 2^2 + \cdots + n^2 = \binom{n+1}{2} + 2\binom{n+1}{3} = \frac{n(n+1)(2n+1)}{6}.$$

Remark

We have found a closed formula for $\sum_{k=1}^{n} k^2$, a highly non-trivial result (where the GAUSS trick would not work)!
Imagine you had been given the problem "Find a closed formula for the sum $\sum_{k=1}^{n} k^2$ ". Would you have got the idea of using the counting problem above to find the answer? No? Don't worry. With practice you will build up a repertoire of ideas, which then helps you to get new ones.

? Problem 5.10

Prove the formula

$$2^n = \sum_{k=0}^{n} \binom{n}{k}$$

($n \in \mathbb{N}$) by counting a set in two ways!

! Solution

What is counted by 2^n? The subsets of $A_n = \{1, \ldots, n\}$, see Problem 2.1 and Section 5.2. And $\binom{n}{k}$ counts the subsets with k elements. So both the left and the right side count the number of subsets of A_n, but on the right we sort them by their number of elements.

More formally: Let $A_n = \{1, \ldots, n\}$ and $X = \mathcal{P}(A_n)$ the power set of A_n. For $k \in \{0, 1, \ldots, n\}$ let X_k be the set of k-element subsets of A_n: $X_k = \{S \subset A_n : |S| = k\}$. Then

$$X = X_0 \cup X_1 \cup \cdots \cup X_n$$

and the X_k are pairwise disjoint. By the addition rule, $|X| = \sum_{k=0}^{n} |X_k|$. The formula now follows from $|X| = 2^n$ and $|X_k| = \binom{n}{k}$. !

You can also interpret the formula in terms of PASCAL's triangle, which was introduced in Exercise E 3.5: The sum of entries in the nth row is 2^n. This can also be proved by induction using the formula in Exercise E 5.14 (do it!), but the counting proof gives more insight.

Here is a proof by counting in two ways which you may know from elementary school:

? Problem 5.11

Prove by counting in two ways that

$$n \cdot m = m \cdot n$$

for all natural numbers n, m.

! Solution

Consider a rectangular array of dots, which is n dots wide and m dots high. If you count the dots column by column then you get $n \cdot m$ dots (each column has m dots, there are n columns); if you count them row by row then you get $m \cdot n$ dots. !

This may look trivial to you – for children who have just learned to multiply it is not.

5.4 Going further: double sums, integrals and infinities

You can find variations on the idea of counting in two ways in all areas of mathematics. Here are some examples. Even if you don't understand all the words in the explanations that follow, keep reading anyway. You will remember the ideas when you learn more mathematics. Then you should also look out for more examples of this principle.

Proofs without words

Figure 5.3 shows that (and why) the identity

$$1 + 3 + \cdots + (2n - 1) = n^2$$

holds, no explanations needed. There are many more pretty proofs without words, just search the internet.

BURNSIDE's lemma

BURNSIDE's lemma is a useful tool for counting structures with symmetry. The triominoes in Problem 5.3 are a good example. The symmetries are the rotations of the triangle by 0, 120 and 240 degrees. BURNSIDE's lemma gives a new way of counting triominoes: For each rotation determine the number of its fixed points, i.e. the number of marked triangles which stay unchanged under this rotation. For the rotation by 0 degrees (i.e. no rotation) that's all triangles (216). For the two other rotations that's only the triangles marked by three equal numbers (6 each). BURNSIDE's lemma tells us that the number

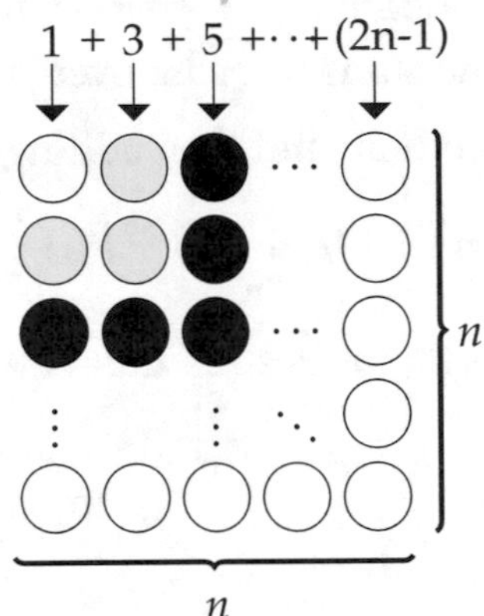

Figure 5.3 A proof without words

of triominoes is precisely the mean value of these three numbers, that is $\frac{1}{3}(216+6+6) = 76$. That is the result we obtained by a different argument.

Why does this work in general? You can find the general statement and a proof of BURNSIDE's lemma in the book (Aigner, 2007), for example. The proof uses counting in two ways. There you can also find Pólya's enumeration theorem, a more sophisticated version of BURNSIDE's lemma, which can be used to count chemical compounds, for example.

Double sums

A variant of counting in two ways is calculating a double sum in two different orders: Suppose you want to find the sum of numbers a_{ij}, $i = 1, \ldots, n$, $j = 1, \ldots, m$ which you can arrange in a rectangular array (matrix) as follows (for $n = 2$, $m = 3$ say):

$$\begin{pmatrix} a_{11} & a_{12} & a_{13} \\ a_{21} & a_{22} & a_{23} \end{pmatrix}.$$

To find the sum of all the numbers you can proceed in (at least) two ways:

- first sum over j for each fixed i, then add the sums (i.e. sum over i): $\sum_{i=1}^{n} \sum_{j=1}^{m} a_{ij}$. This means first summing each row and then adding the row sums, in the example $(a_{11} + a_{12} + a_{13}) + (a_{21} + a_{22} + a_{23})$;

- first sum over i for each fixed j, then add the sums (i.e. sum over j): $\sum_{j=1}^{m}\sum_{i=1}^{n} a_{ij}$. This means first summing each column and then adding the column sums, in the example $(a_{11} + a_{21}) + (a_{12} + a_{22}) + (a_{13} + a_{23})$.

Since we are adding all numbers in both cases, the results must be equal:

$$\sum_{i=1}^{n}\sum_{j=1}^{m} a_{ij} = \sum_{j=1}^{m}\sum_{i=1}^{n} a_{ij}. \tag{5.3}$$

This generalizes counting in two ways. For example, if all $a_{ij} = 1$ then you obtain the counting proof of $n \cdot m = m \cdot n$. Another example is the edge-vertex formula (4.4) for graphs: Suppose a given graph has its vertices numbered by i and its edges numbered by j. Set

$$a_{ij} = \begin{cases} 1 & \text{if edge } j \text{ has vertex } i \text{ as one endpoint} \\ 2 & \text{if edge } j \text{ has vertex } i \text{ as both endpoints} \\ 0 & \text{otherwise.} \end{cases}$$

Then the sum of row i is the degree of vertex i and the sum of column j is 2 for each j. So (5.3) is just the edge-vertex formula (here $n = v$, $m = e$).

Changing the order of summation as in equation (5.3) also gives interesting results for infinite sums (so-called series). A famous example of a series is $\sum_{n=1}^{\infty} \frac{1}{n^2} = 1 + 1/4 + 1/9 + \dots$. There is no simple way to compute the value of this series. A beautiful argument that is based on interchanging summations is the HERGLOTZ trick, see for example (Aigner and Ziegler, 2014). By the way, the value of the series is $\pi^2/6$.

Double integrals

If you are familiar with integration then you know that integrals may be considered as the continuous analogue of sums.

This leads to further generalizations of the idea of counting in two ways: If you have a function of two variables, $f(x, y)$, then you can integrate it first in x and then in y, or first in y and then in x – the fact that the results are the same is known as FUBINI's theorem. If you have several (possibly infinitely many) functions of one variable, $f_1(x), f_2(x), \ldots$, then you can interchange summation and integration: $\sum_i \int f_i(x)\,dx = \int (\sum_i f_i(x))\,dx$.[8]

A different, rather spectacular example of counting in two ways with integrals occurs in the computation of the integral $I = \int_{-\infty}^{\infty} e^{-x^2}\,dx$, the area under the bell-shaped curve e^{-x^2} that appears in probability theory as Gauss's normal distribution. Because you cannot write down an antiderivative for the function e^{-x^2} it is not at all obvious how one should go about computing the integral. Here is an ingenious trick: You write $I = \int_{-\infty}^{\infty} e^{-x^2}\,dx = \int_{-\infty}^{\infty} e^{-y^2}\,dy$ (renaming the integration variable does not change the value of the integral!) and multiply:

$$I^2 = \int_{-\infty}^{\infty} e^{-x^2}\,dx \int_{-\infty}^{\infty} e^{-y^2}\,dy = \int_{-\infty}^{\infty}\int_{-\infty}^{\infty} e^{-(x^2+y^2)}\,dx\,dy$$

This is a double integral, which computes the volume of a certain body B. We use coordinates x, y, z in space. The body B is infinitely extended in the horizontal x,y-directions; its base is the $z = 0$ plane and its top is the graph of the function $z = e^{-(x^2+y^2)}$. The double integral tells us to compute the volume of B 'slice by slice': If you fix y then the points of B having this y-value form a planar slice of B. The area of this slice is precisely the inner integral $\int_{-\infty}^{\infty} e^{-(x^2+y^2)}\,dx$. Then the volume of B is obtained by integrating these areas over y. This was the first 'count', see the left-hand diagram in Figure 5.4.

Now comes the second 'count', see the right-hand diagram in Figure 5.4: Instead of slicing B into planar slices we slice it into cylindrical slices. For $r > 0$ let $C_r = \{(x,y) : x^2 + y^2 = r^2\}$ be the circle of radius r around the origin in the x, y-plane. On C_r the function $e^{-(x^2+y^2)}$ has the constant value e^{-r^2}. So the part of B having $x^2 + y^2 = r^2$ is a cylindrical surface of radius r and height e^{-r^2}. The area of this surface is $2\pi r \cdot e^{-r^2}$. In order to find the volume of B we integrate all these cylinder areas and get

$$\int_0^{\infty} 2\pi r e^{-r^2}\,dr = \pi \int_0^{\infty} e^{-s}\,ds = -\pi e^{-s}\Big|_{s=0}^{\infty} = \pi$$

(substitute $r^2 = s$). Since the two ways of computing the volume (planar or cylindrical slices) must give the same result, we get $I^2 = \pi$

[8] These theorems are only true under certain conditions on f resp. the f_i. See (Pugh, 2010, Section 4.1) or (Apostol, 1967, Vol. I, Section 11.4) for details.

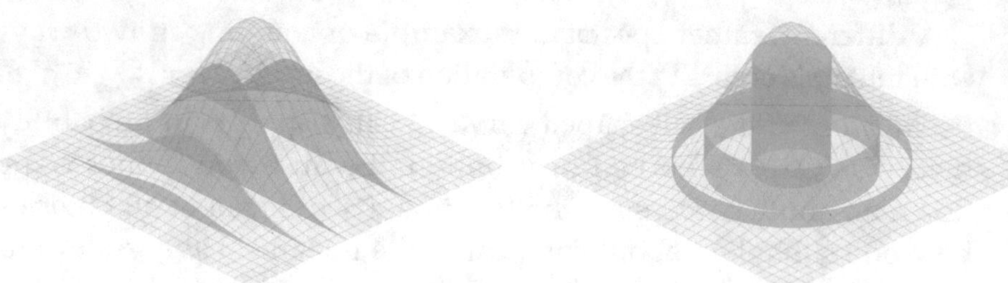

Figure 5.4 Two ways to compute the volume under the graph $z = e^{-(x^2+y^2)}$

and hence $I = \sqrt{\pi}$. So we have derived the surprising result[9]

$$\int_{-\infty}^{\infty} e^{-x^2}\,dx = \sqrt{\pi}.$$

As for the series $\sum_{n=1}^{\infty} \frac{1}{n^2}$ the appearance of π is quite unexpected since there is no sign of anything geometric or 'circle-like' in the integral. Our derivation explains where the π comes from.

Bijections, parties and infinities

The idea of bijections is also ubiquitous in mathematics, for example when considering the question: How big is infinity?

Imagine you throw a big party, and you want to know whether there are as many men as women among your guests. There are (at least) two ways to find out:

1. You count the women and you count the men and compare the numbers.

2. You ask your guests to pair up (always a man and a woman). If no one is left then there are as many women as men.

[9] We omitted one detail in the argument: We used implicitly that for any $r > s > 0$ the circles C_r and C_s have constant distance $r - s$ from each other. If this was not the case we would get an additional factor in the second 'count', stemming from the change of variables formula for integrals. For details see (Pugh, 2010, Section 5.7) or (Apostol, 1967, Vol. 2, Section 11.26).

The second method allows you to ascertain the equality of the numbers without actually knowing the numbers! All you did is establish a bijection between the set of women and the set of men.

The same idea is used for classifying infinities (cardinalities). Two sets are said to be **of the same cardinality** (or equinumerous) if there is a bijection between them. This notion makes just as much sense for infinite as for finite sets since we don't actually need to count the 'number of elements'. However, with infinite sets some funny things happen. For example, the set of natural numbers $\mathbb{N}$ has the same cardinality as the set of even natural numbers, since mapping n to $2n$ for all $n \in \mathbb{N}$ gives a bijection. Even more surprisingly, $\mathbb{N}$ has the same cardinality as the set of rational numbers (fractions), but not as the set of real numbers.[10]

In short: There are as many natural numbers as there are even natural numbers or fractions, but less than there are real numbers.

5.5 Toolbox

The fundamental counting rules (addition rule, multiplication rule, multiple-counting principle) enable us to solve the basic types of counting problems (counting tuples and subsets). You should know the solutions of these problems (the formulas and also their derivations) very well because many counting problems can be reduced to one of them.

Bijections can give a deeper, immediate understanding of counting formulas, for example for the formula $|\mathcal{P}(\{1,\dots,n\})| = 2^n$.

Using the ideas of bijections and counting in two ways one can derive many interesting identities.

Exercises

E 5.1 What is the number of triples (a_1, a_2, a_3) of numbers $a_1, a_2, a_3 \in$ 1
$\{1,2,\dots,10\}$ satisfying $a_1 < a_2 < a_3$?

E 5.2 You throw two dice, one black and one white. How many 1-2
possible outcomes are there? An example of an outcome is 'black 1,

[10]For proofs and more explanations see (Pugh, 2010, Section 1.4), for example.

white 3′. How many outcomes show two different numbers? How many outcomes are there if the dice are indistinguishable?

1-2 E 5.3 In the game of dominoes each tile has two fields, each showing 0, 1, 2, 3, 4, 5 or 6 spots. How many domino tiles are there? Give at least two derivations of your answer.

2 E 5.4 The game of poker is played with a deck of 52 playing cards, showing 13 ranks (Ace, 2, 3, ..., 10, Jack, Queen, King) of 4 suits. A poker hand consists of 5 cards. The value of a hand is determined by its category. The categories are:

- One pair/Three of a kind/Four of a kind: 2/3/4 cards have equal ranks, but apart from this no rank appears multiple times
- Two pair: 2 cards of one rank, 2 cards of another rank, and one card of a third rank
- Full House: 3 cards of one rank and 2 cards of another rank
- Straight: successive ranks
- Flush: all cards have the same suit
- Straight Flush: successive ranks and equal suit

How many poker hands are there? How many of each category are there?

2 E 5.5 How many ways are there to go home from campus, see Figure 5.5? Only direct routes are counted, i.e. those that always go east or north. How many ways are there if you want to buy bread on your way home?

2 E 5.6 Find a formula for the sum of cubes $\sum_{k=1}^{n-1} k^3$ by counting the elements of the set

$$\{(a,b,c,d) : a,b,c,d \in \{1,\ldots,n\},\ a < d,\ b < d,\ c < d\}$$

in two ways.

2 E 5.7 Let $P_1,\ldots,P_n$ be points on a circle. Assume that no three chords (i.e. lines P_jP_k) intersect in a point. Find a bijection between the set of 4-element subsets of $\{P_1,\ldots,P_n\}$ and the set of intersection points of chords.

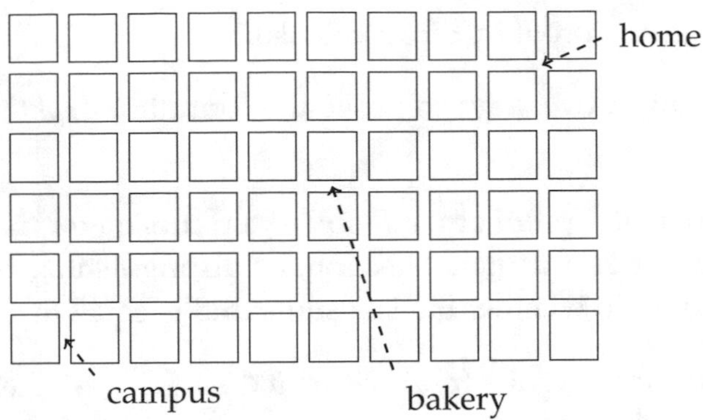

Figure 5.5 Paths in a street grid

How many intersection points are there?

E 5.8 26 people named $A, B, C, \ldots, Z$ write, one after another, their birthday in a calendar of the year 2016, a leap year. Each person writes her name next to the date of her birthday. How many possible calendars are there? In how many of them does everyone have a different birthday? In how many do at least two people have the same birthday? In how many does at least one person have her birthday on January 1? In how many exactly one? In how many exactly two? In how many does exactly one person have her birthday on January 1 and exactly one on March 28? 2

Also answer these questions in the case where the people write a tick mark next to the date of their birthday, instead of their name.

E 5.9 Let $f_1, f_2, f_3, \ldots$ be the FIBONACCI numbers: $f_1 = 1$, $f_2 = 2$ and $f_n = f_{n-1} + f_{n-2}$ for all $n \geq 3$. Prove the formula $f_{2n} = f_n^2 + f_{n-1}^2$ for all $n \geq 2$. Give two proofs: one using induction and one by counting in two ways. Here you may use that f_n is the number of trains of length n with coaches of length 1 or 2, see Exercise E 2.8. Find more formulas for the FIBONACCI numbers, e.g. a similar one involving f_{2n+1}. 2

E 5.10 Prove the equation $4 \cdot \binom{50}{4} = 50 \cdot \binom{49}{3}$ by counting in two ways: how many ways can you choose a group of 4, with a leader, from 50 people? Generalize this identity to any number of people and group 2 sizes. Also give a proof by direct calculation.

2 E 5.11 Count in two ways to prove the formula $\sum_{k=0}^{n} k\binom{n}{k} = n2^{n-1}$ for $n \in \mathbb{N}$.

2 E 5.12 Give a new proof of the FIBONACCI recursion for the numbers o_n in Exercise E 2.15 using a bijection, by distinguishing representations according to whether the last summand is equal to 1 or not.

2-3 E 5.13 Consider the pairs (a, b) with $a, b \in \{0, 1, \ldots, n\}$. Let e_n be the number of such pairs for which $a + b$ is even and o_n the number of pairs for which $a + b$ is odd. What do you observe for $n = 0, 1, 2, 3$? State a conjecture and give proofs via bijection, direct calculation and induction. Can you generalise to more than two numbers?

2 E 5.14 The binomial coefficients satisfy the recursion

$$\binom{n+1}{k} = \binom{n}{k-1} + \binom{n}{k}.$$

Give two proofs: one by direct calculation, the other by counting the k-element subsets of $\{1, \ldots, n+1\}$ in two ways. Conclude that $\binom{n}{k}$ is the kth entry in the nth row of PASCAL's triangle, see Exercise E 3.5. Here you always start counting at zero.

2 E 5.15 The binomial coefficients appear in many places in mathematics. They get their name from the **general binomial theorem**:

$$(a+b)^n = a^n + \binom{n}{1}a^{n-1}b + \binom{n}{2}a^{n-2}b^2 + \cdots + \binom{n}{n-1}ab^{n-1} + b^n$$

for real numbers a, b and $n \in \mathbb{N}_0$. Check the formula for $n = 2$ and $n = 3$ by multiplying out the left-hand side. Give two proofs for the general formula: one by induction using the formula in Exercise E 5.14, the other by using the meaning of $\binom{n}{k}$ as number of subsets.

2 E 5.16

a) Check by direct calculation that the formula (2.9) for the nth FIBONACCI number always yields a rational number. That is, check that when multiplying out the powers using the general binomial theorem (see Exercise E 5.15) all square roots disappear.

b) We know from their recurrence relation that the FIBONACCI numbers are integers. This implies something about the divisibility of certain combinations of binomial coefficients. What can we conclude?

E 5.17 Let $a = 1 + \sqrt{2}$. Calculate the decimal representations of $a, a^2, a^3, a^4, \ldots$ with your calculator. What do you observe? Find a reason. What is the digit in the 50th decimal place of a^{200}? 2-3

E 5.18 Find at least one more proof for the equality of the numbers of even and odd subsets of $\{1, \ldots, n\}$. 2-3

E 5.19 Let $n \in \mathbb{N}$. Find a bijection proof for the fact that the number of ways to write n as an ordered sum of natural numbers is 2^{n-1} (see Exercise E 1.9). 3

E 5.20 Let $n \in \mathbb{N}$. How many solutions does the equation $a + b + c = n$ have with $a, b, c \in \mathbb{N}$? 3

E 5.21 Use the following idea to give a new derivation of the formula for the number of regions which result from n lines in the plane in general position (Problem 1.3): 2-3

First turn the plane so that no line is horizontal. Then associate with every region its lowest point, if it has a lowest point. Do all intersection points of lines arise as lowest points? How many regions do not have a lowest point?

E 5.22 Generalise the idea of Exercise E 5.21 to find the number of regions into which space is divided by n planes in general position (no two parallel, no three through a line, no four through a point). 3

E 5.23 Generalise the idea of Problem 5.8 to the counting of triangles that can be formed from the n points, and derive an identity. 2-3

E 5.24 Find the number of regions into which the interior of a circle is divided if you choose n points on its circumference and draw all chords between them. We assume the points are chosen so that no three of the chords pass through a common point. 3-4

Find the number for $n = 2, 3, 4, 5$ and state a conjecture. Then find the number for $n = 6$. Find a general formula.

6 General problem solving strategies: Similar problems, working forward and backward, interim goals

General problem-solving strategies are strategies which can be used in everyday life, not just in mathematics: If I want to solve a problem then it will help me to recall how I solved a similar problem. If I want to reach a goal then I can think about which steps I should do first in order to get there (working forward); or I can think about what could be the last step, reaching the goal (working backward), and what interim goals I could set for myself.

We now take a systematic look at these problem-solving strategies. We also represent them in schematic diagrams, which can help us to keep track of our progress. Then we investigate two problems, one from geometry and one about representing integers as sums, and pay particular attention to the strategies we use to solve them.

6.1 General problem solving strategies

The first step in problem-solving is to **understand the problem.** We ask ourselves:

- ❑ **What is given?**
- ❑ **What are we looking for?**

In most mathematical problems either you want to find something, or you want to give a proof. For example, in the log-cutting problem 1.1 and in most counting problems you want to find a number, and in Problem 4.1 on EULER's formula you are asked to give a proof.

In a **finding problem** you are given *data* (for example, the length of the log) and you are looking for certain *unknowns* (for example the time to cut it into pieces).

In a **proving problem** you are given *conditions* (for example, the graph G is plane and connected) and you are looking for the *proof of a conclusion* (for example, EULER's formula holds for G).

D. Grieser, *Exploring Mathematics*, Springer Undergraduate Mathematics Series, https://doi.org/10.1007/978-3-319-90321-7_6

Not every mathematical problem can be classified as a finding problem or a proving problem. In a finding problem you will also need to prove that the method you used is correct. However, often this proof is already implicit in the derivation of the solution. In addition, there are many mathematical problems where you just want to understand some situation, and where it is not clear at the outset what you want to find or prove. You will find an example of this in Chapter 9.3.

In order to solve the problem we need to find a connection from the data to the unknown, from the conditions to the conclusion.

Figure 6.1 The problem: Looking for a connection

The most important general strategies

- ❑ **Have I seen a similar problem before?**

  This is one of the first questions that you should ask yourself.

- ❑ **Working forward: How can I use what is given?**
  Working backward: How can I reach the goal?

  You can approach a problem from the front or from the back. Approaching it from the front means asking how you can use the data/the conditions. Approaching it from the back means focusing on the goal and asking how you could get there. See Figures 6.2 and 6.3.

- ❑ **Can I formulate useful interim goals?**
  An interim goal can be anything that helps to bridge the gap between the given and the goal. See Figure 6.4.

You already know some other general strategies: looking at special cases and examples, making a sketch or a table, simplifying, introducing suitable notation, formulating a conjecture. Typically you use a combination of these (and other) strategies. **If one strategy**

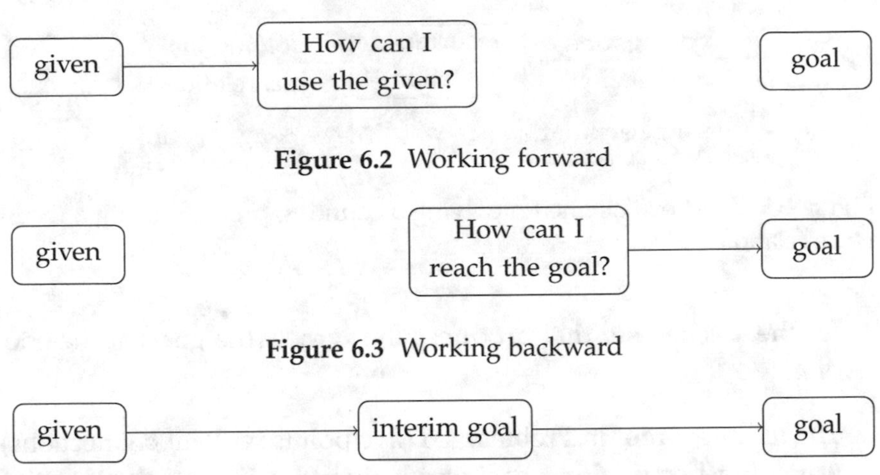

Figure 6.2 Working forward

Figure 6.3 Working backward

Figure 6.4 Solution scheme with an interim goal

turns out not to help then try another. For example, you first try working forward, and if you don't see what to do, you try working backward. And when this has given you more understanding, you try working forward again. Sometimes you also have more complex solution schemes with several interim goals. For example, we could first formulate one interim goal and then, considering how we could reach that interim goal, formulate another interim goal.

Let us look back at some of our solutions from previous chapters from this perspective.

- In the log cutting problem 1.1 we set the interim goal of finding the number of pieces. This number was easy to find from the data (the length of the log), and from there it was easy to determine the time.

- We worked forward when solving the FIBONACCI recursion in Chapter 2.4, and also in most counting problems in Chapter 5.

- In counting problems the search for a recurrence relation can be a useful interim goal. Once we set ourselves this interim goal we can look for ways to derive a recurrence relation; this is working backward (from the interim goal to the data). See the problems in Chapter 2.

- In Problem 1.2 (zeroes of 100!) we worked backward: we looked

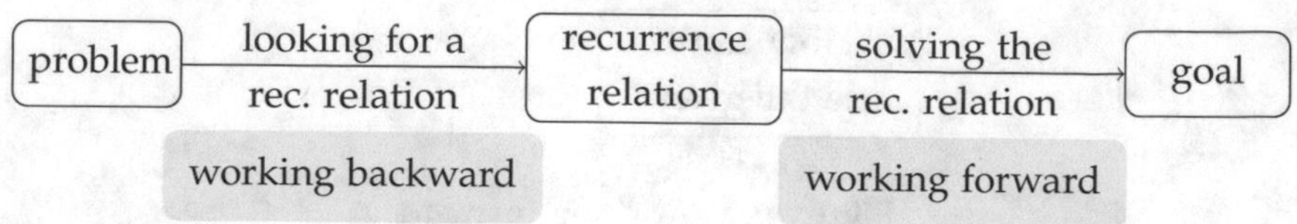

Figure 6.5 Solution scheme for solving a counting problem using a recurrence relation

for the mechanism that produces the zeroes (the goal) at the end of $n!$.

- The indirect proof in Problem 4.6 (five points with all connections) was a kind of working backward: Suppose the conclusion (the goal) is wrong, what follows from this?

A simple example of working backward

? Problem 6.1

A merchant has a crate of apples. A customer comes and buys half of the apples and one extra. Then another customer comes and buys half of the remaining apples and one extra. Then a third and a fourth and a fifth customer come and do the same. At the end there is one apple left. How many apples did the merchant have at the beginning?

! Solution

The simplest way to solve this is to start at the end: At the end the merchant has one apple, so before giving the extra apple away he had 2, and before giving away the last half he had 4. So he had 4 apples before the fifth customer came. Similarly he had 5 apples before the fourth customer got the extra apple and before that 10. So the merchant had 10 apples before the fourth customer came. Similarly, $22 = (10 + 1) \cdot 2$ apples before the third, $46 = (22 + 1) \cdot 2$ before the second and $94 = (46 + 1) \cdot 2$ before the first customer. Therefore the merchant had 94 apples at the beginning. !

This is a classic example of working backward in a finding problem. Note that doing "first halve, then subtract one" backward means "first add one, then double". When turning things around, the order also gets turned around. See also Exercise E 6.12.

6.2 The diagonal of a cuboid

We now investigate a geometric finding problem.

? Problem 6.2

Find the length d of the diagonal of a cuboid with side lengths a, b, c.

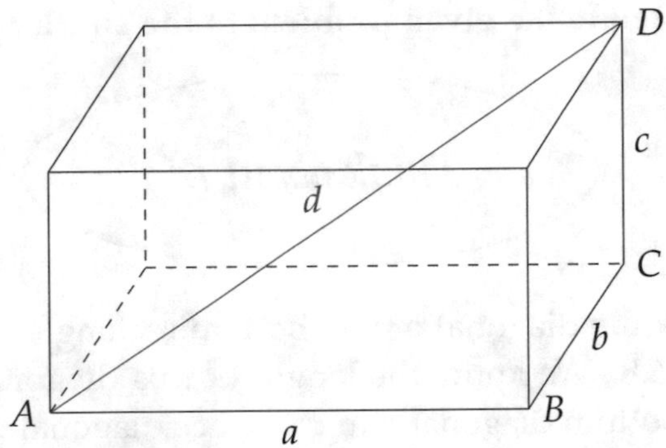

Figure 6.6 Diagonal of a cuboid

Investigation and solution

▷ We first make a **sketch**, see Figure 6.6.

▷ **What is the data?** The side lengths a, b, c. **What is the unknown?** The length d of the diagonal $\overline{AD}$.

▷ Special cases and examples seem to be useless here.

▷ **Have I seen a similar problem before? Can I simplify the problem?**

Solid geometry is harder than plane geometry. Maybe we can solve a similar problem in plane geometry first? Here is such a problem: Find the length x of the diagonal of a rectangle with side lengths a, b. See Figure 6.7.

You know this from school: The triangle ABC is right-angled, so we have $x^2 = a^2 + b^2$ by PYTHAGORAS' theorem, hence $x = \sqrt{a^2 + b^2}$.

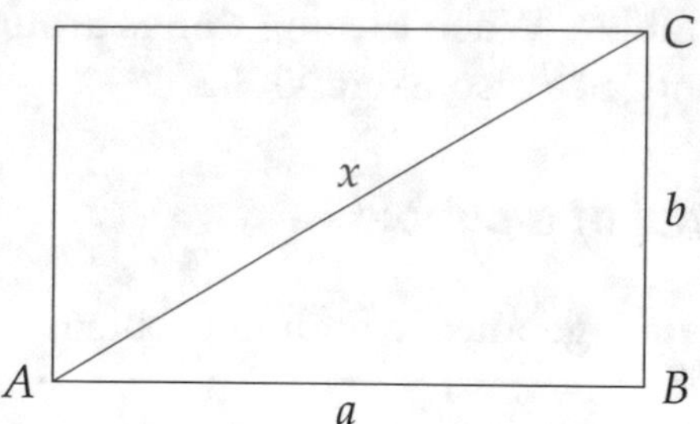

Figure 6.7 Diagonal of a rectangle

▷ **How can I relate the given problem to the similar problem?**

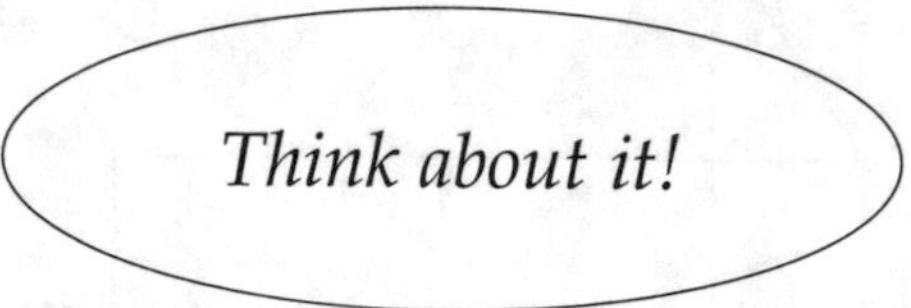

▷ Let us draw the diagonal of the 'bottom rectangle' of our cuboid, see Figure 6.8. We know the length of this diagonal, x. Can we relate this bottom diagonal x to the space diagonal d?

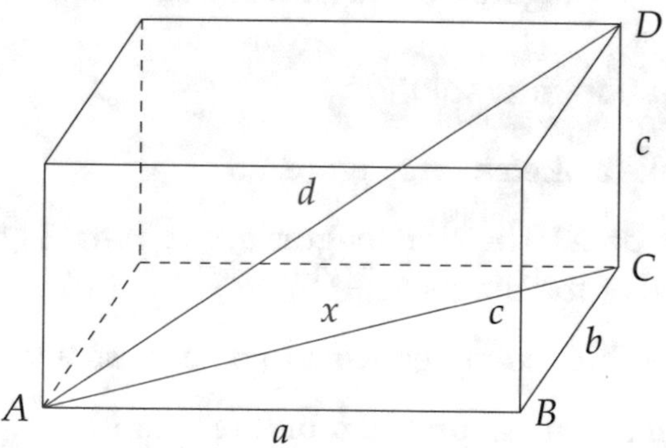

Figure 6.8 Sketch for the solution

▷ *Observation:* The bottom diagonal and the space diagonal are two sides of the triangle ACD. Let us take a close look at this triangle. Its third side length is c. Do we know anything else? The angle at C must be a right angle since the side $\overline{CD}$ is orthogonal to the bottom plane, hence to any line contained in the bottom plane.

▷ Therefore we can use PYTHAGORAS' theorem again, now for the triangle ACD. We get $d^2 = x^2 + c^2$. Using $x^2 = a^2 + b^2$ we get $d^2 = a^2 + b^2 + c^2$, hence

$$d = \sqrt{a^2 + b^2 + c^2}.$$

This is the solution!

Review of Problem 6.2

We were given a, b, c, the unknown was d. The situation is shown schematically in the left part of Figure 6.9. The question mark symbolizes the connection that we are looking for. The key to the solution was to look at the bottom diagonal. It acts as a **link** between the given and the unknown. The right part of Figure 6.9 shows the solution scheme.[1]

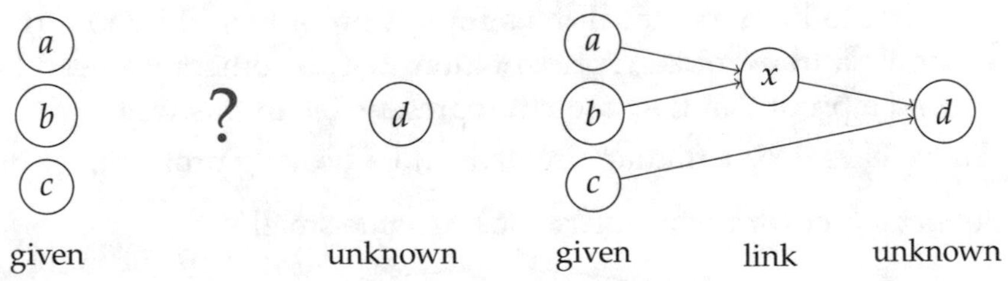

Figure 6.9 The problem and the solution scheme

6.3 The problem of trapezoidal numbers

Let us try a more difficult problem!

Problem 6.3

Which natural numbers n can be written as the sum of several consecutive natural numbers?

[1] George Pólya discusses this example at length in his classic 'How to solve it – a new aspect of mathematical method', (Pólya, 2014), in a hypothetical teacher - student interaction. Delightful!

For brevity we call such a representation of n a **representation as a trapezoidal number,** or a trapezoidal representation. Look at Figure 6.10 to understand why.

Figure 6.10 The trapezoidal numbers $12 = 3+4+5$ and $11 = 5+6$

Investigation

▷ **What is given?**

Only the definition of a trapezoidal number.

What are we looking for?

1. The answer to the question: Which numbers n have trapezoidal representations?
2. A proof that our answer is correct. For the numbers that we claim to be trapezoidal the simplest proof would be to write down a trapezoidal representation. For the others we need to find a proof that they are not representable in this way.

So we have both a finding problem and a proving problem.

▷ To **get a feel for the problem** look at some small n.

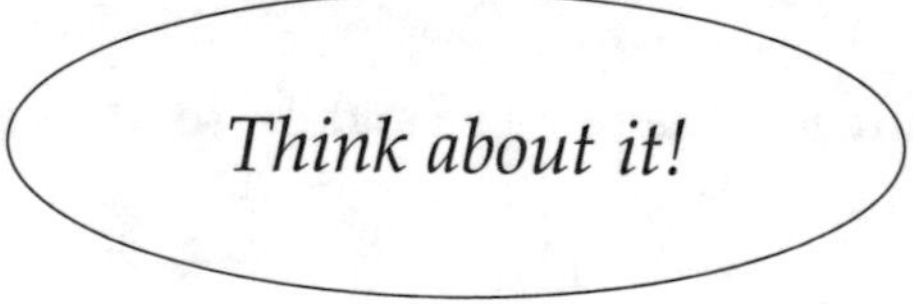

You get Table 6.1. The numbers 1, 2, 4, 8 are not trapezoidal. This may give you an idea:

> *Vague conjecture:* All numbers except the powers of 2 are trapezoidal.

The conjecture is quite vague because we have only a little data, and because there does not seem to be a good reason why powers of 2 should play a role for trapezoidal numbers. If you extend the table then the conjecture holds up, but you cannot go on forever. **We need a general method.**

The table also shows that a number can have several trapezoidal

n	trapezoidal representations
1	none
2	none
3	$1+2$
4	none
5	$2+3$
6	$1+2+3$
7	$3+4$
8	none
9	$4+5$ or $2+3+4$

Table 6.1 Trapezoidal representations

representations. But this is not part of the question. See Problem E 6.5, however.

▷ How can we approach the conjecture? Looking at the table, you might recognize a pattern for odd n.

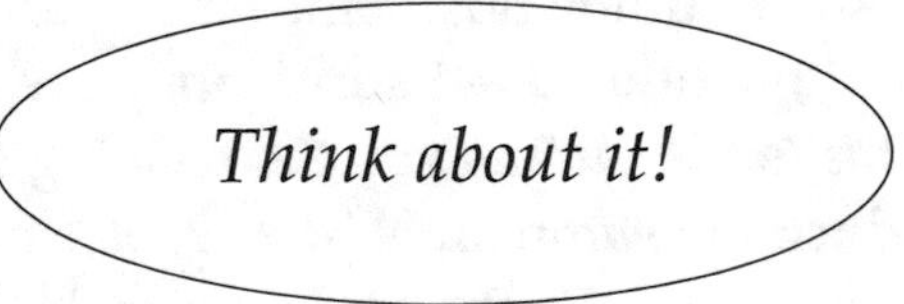

▷ If $n > 1$ is odd then it can be written as the sum of two consecutive integers: Write n as $n = 2m + 1$ where $m \in \mathbb{N}$, then $n = m + (m + 1)$ is the representation we needed.

So odd numbers bigger than one are trapezoidal.

▷ To prove the general (vague) conjecture we need to do two things:[2]

1. Find a trapezoidal representation for each n which is not a power of 2.

2. Prove that powers of 2 do not have such a representation.

We start by looking at the first task. Sometimes it helps to **reformulate:** What does it mean that n is not a power of 2?

It means that n has an odd divisor (other than 1). For example, 12

[2]However, we must always remember that it is only a conjecture. It might happen that during the investigation we find out that it is wrong. Keep your eyes open.

is divisible by 3 and therefore cannot be a power of 2. And any power of 2 cannot be divisible by any odd number other than 1.[3] So apparently odd divisors play a role.

▷ *First attempt:* Let us try to understand in an **example** why odd divisors could be useful: $50 = 5 \cdot 10$. *Idea:* We could write this as $50 = 10 + 10 + 10 + 10 + 10$, then reduce the first two tens by 2 and 1 and enlarge the last two tens by 1 and 2 – this will not change the sum, and we get $50 = 8 + 9 + 10 + 11 + 12$, a trapezoidal representation of 50.

▷ This looks promising! You can see that this works for any odd number of summands, but not for an even number: The same method would transform $20 = 10 + 10$ into $20 = 9 + 11$, but here 9 and 11 are not consecutive integers. So odd numbers play a role here, we are on the right track. **Can we formulate this procedure in general?**

We let the example guide us. We write $n = xy$ with odd $x > 1$. In the example we had $x = 5$, $y = 10$. We write $n = y + \cdots + y$ where there are x summands. Since x is odd there is a central summand, and the other $x - 1$ summands are on the left and on the right, so that's $\frac{x-1}{2}$ summands each. Going to the left from the center we reduce the summands by 1, 2, etc., and to the right we increase them by 1,2 etc. So the last summand will be reduced resp. increased by $\frac{x-1}{2}$. We obtain

$$n = (y - \frac{x-1}{2}) + \cdots + y + \cdots + (y + \frac{x-1}{2}),$$

where each summand is one bigger than the preceding summand. This is the desired trapezoidal representation of n. Is it really?

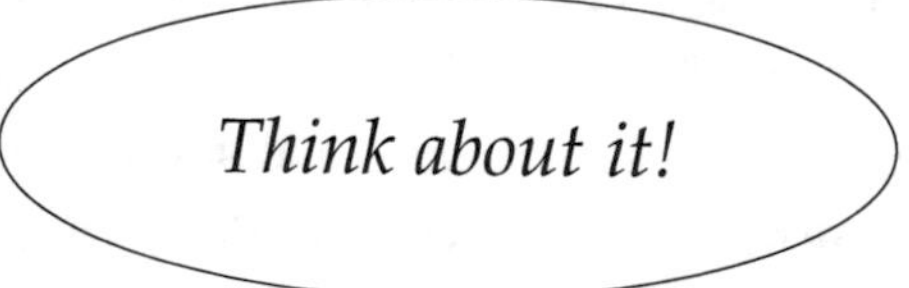

Look again! The summands must be not only consecutive but also positive. This is only the case if $y - \frac{x-1}{2}$ is positive. For example, if we start with $14 = 7 \cdot 2$ then $x = 7$, $y = 2$, so $y - \frac{x-1}{2} = -1$ and our

[3] See Chapter 8 for more on divisibility and divisors.

procedure yields $14 = (-1) + 0 + 1 + 2 + 3 + 4 + 5$. This is correct, but not what we want. So our method only yields a trapezoidal representation if $y > \frac{x-1}{2}$.

There is a nice little trick to turn this forbidden representation into a permitted one. Do you see it? See Exercise E 6.4. For now we will start again and try to approach our problem in a more systematic way. This will also give us a way to find out *how many* representations n has, see Exercise E 6.5.

▷ *Second attempt:* We focus on our goal **(working backward).** What are we looking for? A representation

$$n = a + (a+1) + \cdots + b \quad \text{where } a < b$$

and a, b are natural numbers. It would be nice if we could simplify ('calculate') the right hand side. How to do that?

▷ **Have I seen a similar problem before?** Yes, when calculating $1 + \cdots + n$. Remember the pretty GAUSS trick from Section 1.3? Let us apply it here:

$$\begin{array}{ccccccc} n & = & a & + & \dots & + & b \\ n & = & b & + & \dots & + & a \\ \hline 2n & = & (a+b) & + & \dots & + & (a+b) \end{array}$$

How many summands $a + b$ are there? You can read it off from the first line: there must be $b - a + 1$ summands (numbers from a to b, counting both a and b, therefore $\cdots + 1$). We get

$$2n = (a+b)(b-a+1)\,. \tag{6.1}$$

▷ What are we looking for? For given n we want to find a, b solving Equation (6.1). This equation looks complicated. **Can we simplify it?** We could **introduce some notation:**

$$c = a + b, \quad d = b - a + 1\,. \tag{6.2}$$

Thus $2n = cd$, which looks much nicer. How can we get a, b from c, d? We just need to solve the system of equations (6.2) for a and b: adding the equations yields $c + d = 2b + 1$, subtracting yields $c - d = 2a - 1$. So we get

$$a = \frac{c-d+1}{2}, \quad b = \frac{c+d-1}{2}\,. \tag{6.3}$$

▷ What have we accomplished? We have replaced the complicated task of solving Equation (6.1) (given n, find solutions a, b) by two simpler tasks:

a) Find c, d satisfying $2n = cd$.

b) From c, d determine a, b using Equations (6.3).

This is an example of an **interim goal:** to find c, d.

▷ Let us look at the new tasks. Part b) is easy, just plug in. Part a) is also easy. For example, one could take $c = 2, d = n$ or vice versa. Are we done yet? Something must be missing since this works for all n, including powers of 2.

Let us take a closer look. What are the conditions on a, b? In order for $n = a + (a+1) + \cdots + b$ to be a trapezoidal representation, a, b must satisfy some conditions:

- $a, b \in \mathbb{N}$. The equations (6.3) show that then $c - d + 1$ and $c + d - 1$ must be even, and $c > d$. We get:

$$c, d \text{ have different parity}^4\text{, and } c > d. \tag{6.4}$$

  Conversely, it is clear that this guarantees $a, b \in \mathbb{N}$.

- $a < b$. Let us rewrite this:

$$\begin{aligned} a < b &\iff \frac{c-d+1}{2} < \frac{c+d-1}{2} \\ &\iff 0 < \frac{c+d-1}{2} - \frac{c-d+1}{2} \\ &\iff 0 < d - 1 \iff 1 < d\,. \end{aligned}$$

  We see that $a < b$ is equivalent to

$$d > 1\,. \tag{6.5}$$

▷ What did we accomplish? We obtain a trapezoidal representation of n if and only if we can write $2n = cd$ with c, d satisfying the conditions (6.4) and (6.5).

▷ What does this mean for n? The conditions are: $c > d > 1$, and c, d have different parity. So one of the numbers c, d must be odd,

[4] Parity is the property of an integer being even or odd. So one of the numbers c, d must be even, the other odd.

hence n must have an odd divisor, hence it cannot be a power of 2. Conversely, any odd divisor of n yields c and d. So n has a trapezoidal representation if and only if it is not a power of 2.

Let us write up our solution cleanly and fill in some details.

Solution of Problem 6.3

Claim: n has a trapezoidal representation if and only if n is not a power of 2.
Proof: We prove this in several steps. More precisely, we prove that the following statements are all equivalent:

(i) n has a trapezoidal representation.

(ii) There are $a, b \in \mathbb{N}$, $a < b$ satisfying $2n = (a+b)(b-a+1)$.

(iii) There are $c, d \in \mathbb{N}$ of different parity satisfying $c > d > 1$ and $2n = cd$.

(iv) n is not a power of 2.

Proof of (i) $\iff$ (ii): By definition n has a trapezoidal representation if and only if there are $a, b \in \mathbb{N}$, $a < b$ satisfying $n = a + \cdots + b$. By the calculation above, $a + \cdots + b = \frac{1}{2}(a+b)(b-a+1)$. Therefore n has a trapezoidal representation if and only if there are $a, b \in \mathbb{N}$, $a < b$ satisfying $2n = (a+b)(b-a+1)$.

Proof of (ii) $\Rightarrow$ (iii): Suppose there are a, b as in (ii). Let $c = a + b$, $d = b - a + 1$. Then $2n = cd$. Also, we get (6.3) by the calculation there, and this implies (6.4) and (6.5) as shown there. These are the conditions on c, d in (iii).

Proof of (iii) $\Rightarrow$ (ii): Suppose there are c, d as in (iii). Define a, b by (6.3). Then $c = a + b$, $d = b - a + 1$, so $2n = (a+b)(b-a+1)$, and $a, b \in \mathbb{N}$, $a < b$ follow from (6.4) and (6.5) as proved there.

Proof of (iii) $\Rightarrow$ (iv): Suppose (iii) holds. Since c, d have different parity, one of them is odd. Since both are bigger than 1 it follows that $2n$ must have an odd divisor. Then n must also have an odd divisor, so n cannot be a power of 2.

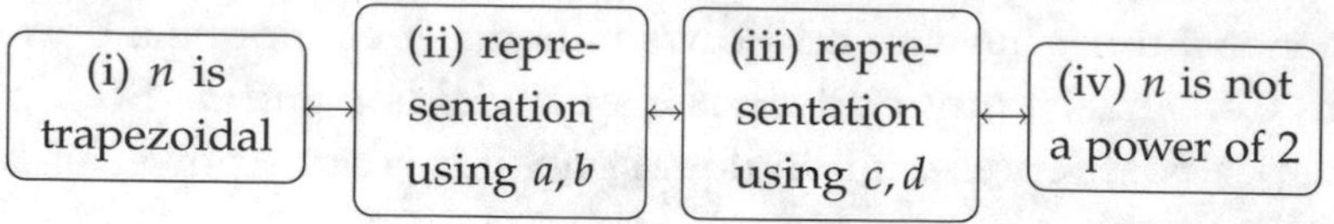

Figure 6.11 Solution scheme for the trapezoidal number problem

Proof of (iv) ⇒ (iii): Suppose (iv) is true. Since n is not a power of 2 it must have an odd divisor $x > 1$. Let $y = \frac{2n}{x} = 2 \cdot \frac{n}{x}$. Then $2n = xy$. Since x is a divisor of n the number $\frac{n}{x}$ is a natural number. Therefore, y is even, in particular $y > 1$ and x, y are different. So if we set

$$c = \max\{x, y\}, \quad d = \min\{x, y\},$$

(that is, c is the bigger and d the smaller of the numbers x, y) then c, d satisfy the conditions of (iii).

All in all, we obtain that (i) and (iv) are equivalent, as required. !

Review of Problem 6.3

- The first attempt was based on an observation about examples and generalizing this. In the second attempt we started by formulating the goal as an equation; working backward and simplifying we were able to formulate interim goals that finally led us to the solution.
- One challenge in complex problems and long investigations is not to get lost in the details, **to keep track of where we are in the solution process.** Therefore it is important to ask yourself again and again: what is given, what are we looking for, what have we accomplished so far? It is a good idea to make notes of that during the investigation.

  Writing things up in a tidy, ordered manner is an important check on whether the argument is complete.
- In the solution you saw a clear, systematic way to write up arguments with several intermediate steps.
- The scheme of our solution is given in Figure 6.11.

6.4 Going further: sum representations of integers

Trapezoidal numbers, triangular numbers etc.

There are also other kinds of 'figurate numbers': maybe you have heard of *triangular numbers*. These are numbers which occur as numbers of points in certain triangular patterns, see Figure 6.12. The nth triangular number is $1+2+\cdots+n$. You already know a formula for this. Square numbers are well-known. From pentagons you get the pentagonal numbers (look them up on Wikipedia!). Very surprisingly they come up in the *partition number problem:* in how many ways can you write a natural number n as a sum of natural numbers in nonincreasing order? For example $4 = 3+1 = 2+2 = 2+1+1 = 1+1+1+1$, that's five ways for $n = 4$; here $3+1$ and $1+3$ are considered to be the same representation. This problem is much more difficult than the problem where $3+1$ and $1+3$ are counted as different, see Exercises E 1.9 and E 5.19.[5]

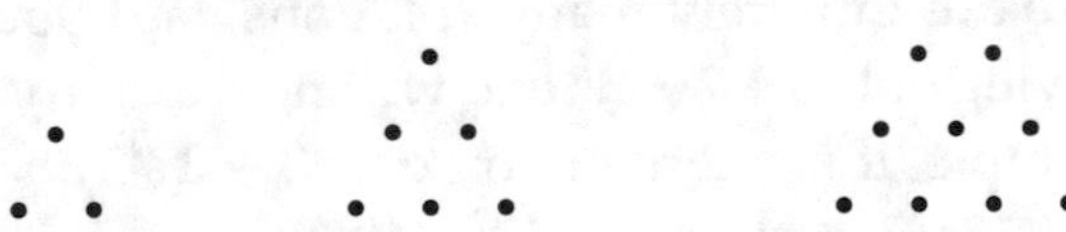

Figure 6.12 The triangular numbers 1, 3, 6, 10

Sum representations of natural numbers

Questions such as "Which natural numbers can be represented in a certain way?", "... and in how many ways?" are a rich source of mathematical problems, which have led to many important developments. The difficulty of the problems varies drastically with the kind of representation. Here are some more examples. You can find solutions for some of them in the books (Hardy and Wright, 2008), (Silverman, 2012).

- $n = p_1 + p_2$: Representing n as the sum of two primes. Most mathematicians believe that every even number n can be represented in this way. This conjecture was first formulated by GOLDBACH

[5]See (Hardy and Wright, 2008), and also 'Partition' and 'Pentagonal number theorem' in Wikipedia for explanations and references.

in 1742. The conjecture is known to be true for all even n up to 4×10^{18}, but is it true for all n? This is still unknown, in spite of intensive efforts by many mathematicians.

- $n = x^2 - y^2$: Representing n as the difference of two squares. (From now on we always assume $x, y \in \mathbb{N}$.) You can solve this, see Exercise E 6.6. When modified slightly it gets much more difficult:
- $n = x^2 + y^2$: Representing n as the sum of two squares. This is a very pretty but difficult topic. For example, it can be shown that all prime numbers that leave the remainder 1 when divided by 4 can be represented in this way (for example, $13 = 2^2 + 3^2$), but those that leave the remainder 3 cannot (for example, 19). See also Exercise E 8.10. The answer for general n is known but a little more complicated.
- $n = x^2 - 2y^2$: Representing n as the difference of a square and twice a square. This is also not easy, but known. The equation $x^2 - 2y^2 = 1$ is called PELL's equation. One solution is $x = 3, y = 2$. There are infinitely many solutions, and you obtain them by multiplying out $(3+2\sqrt{2})^k$ and writing it as $x + y\sqrt{2}$ with $x, y \in \mathbb{N}$. For example, if $k = 2$ then you get $(3+2\sqrt{2})^2 = 9 + 2 \cdot 3 \cdot 2\sqrt{2} + 8 = 17 + 12\sqrt{2}$, and indeed $17^2 - 2 \cdot 12^2 = 1$. It is a nice exercise to check that (x, y) is a solution of PELL's equation for each k. It is a little harder to show that you get all solutions in this way.
- $n = x^2 + y^2 + z^2 + u^2$, $x, y, z, u \in \mathbb{N}_0$: Representing n as the sum of at most four squares. This is possible for all n (LAGRANGE's four-square theorem), but again this is not easy to prove.
- $n = 2^{i_0} + 2^{i_1} + \cdots + 2^{i_k}$, $i_0 < \cdots < i_k$, k arbitrary: Representing n as the sum of different powers of 2. For each n there is a unique such representation (exercise!). This is a restatement of the fact that one can represent numbers in the binary system, hence is the basis of all computer technology.

Exercises

1 **E 6.1** Find perimeter and area of the shape in Figure 6.13 (not by measuring, it is not drawn to scale). Which strategies do you use?

1-2 **E 6.2** You have two pots, with volumes 9 and 4 litres respectively.

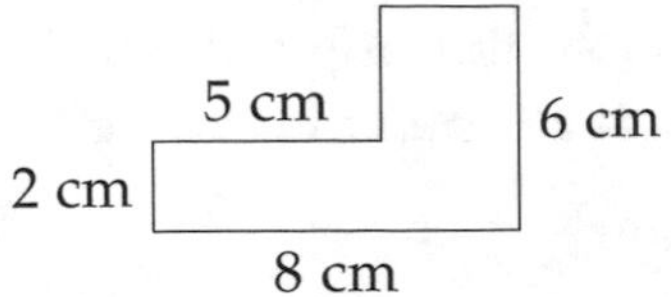

Figure 6.13 Exercise E 6.1

How can you measure precisely 6 litres? You have no tools except for the pots and an unlimited supply of water. Formulate a suitable interim goal. Is it useful to work backward?

E 6.3 Problem 6.2 (diagonal of a cuboid) is symmetric with respect to the side lengths a, b, c: each of them plays the same role. Is the solution formula symmetric? Is the derivation symmetric? If not, what can you say about alternative derivations? 2

E 6.4 Look back at the investigation of Problem 6.3. Complete the first attempt at proving that a trapezoidal representation of n exists if n is not a power of two. 2

E 6.5 How many representations of the numbers 12 and 750 as trapezoidal numbers are there? How many ways are there for an arbitrary $n \in \mathbb{N}$? 2-3

E 6.6 Which numbers $n \in \mathbb{N}$ can be written as the difference of two square numbers? 2

E 6.7 Starting at 1 add the natural numbers in increasing order one by one. Is it possible to obtain a 5 digit number with all digits equal? 2-3

E 6.8 Let digits $a_1, \ldots, a_n$ be given. Prove that there is a natural number N such that the decimal representation of $\sqrt{N}$ has the digits $a_1 \ldots a_n$ immediately after the decimal point. 3

E 6.9 Let three points A, B, C in the plane be given. Explain a constructive procedure to find a line that passes through A and has the same distance from B and C. 2-3

E 6.10 Write 4 zeroes and 5 ones round a circle in any order. Keep repeating the following step: Consider each pair of neighbouring numbers. If they are equal, write a new one between them; if different, 2 a new zero. Now delete the original (so that you are left with 9 numbers again). Can it happen that you ever get 9 ones? Can you ever get 9 zeroes?

3-4 **E 6.11** 100 cards lie in a row on the table, face down. First you turn over the cards at positions 2,4,6,... Then you turn over the cards at positions 3,6,9,... (so that now the card at position 6 is face down again), then the cards at positions 4,8,12,... are turned etc. until in the last round only the card at position 100 is turned over. Which cards lie face down at the end?

2 **E 6.12** What is the solution of Problem 6.1 if there are n customers instead of 5?

2 **E 6.13** Investigate whether the number $x = \sqrt{4+\sqrt{7}} - \sqrt{4-\sqrt{7}} - \sqrt{2}$ is positive, negative or zero.

7 Logic and proofs

We formulate mathematical arguments in our natural language. But our everyday language is often ambiguous, and quite often you hear logical fallacies. Therefore, if you want to argue reliably then you should know well both the basic logical structures and the phrases we use to express them. These are the topic of the first section in this chapter.

In the course of a mathematical investigation you make observations, discover patterns, have insights, make conjectures. In order to be sure that a conjecture is true you need a *proof*.

Logic is one of the foundations of proving. It helps you to be sure that your argument is correct. How to *find* a proof is quite a different matter. Experience will help, and it is useful to have a good command of problem solving strategies, and to know the main *types* and *patterns of proofs*.

The general types of proof are direct proof, indirect proof, and proof by contradiction. Proof patterns are more specific to context. In the second section of this chapter you will learn about the general types of proofs and about some patterns which are common in proofs of formulas, proofs of existence, and proofs of non-existence or impossibility. In subsequent chapters you will study three of these patterns in greater depth.

Of course there are no limits to your creativity in finding your own ways of proving.

7.1 Logic

When doing mathematics we try to understand, calculate, determine things. We formulate our conjectures or results as **propositions.** In logic we investigate how propositions are combined to form new propositions and how the truth value (true or false) of the new propositions depends on that of the old ones. We also study how to draw correct conclusions from true propositions.

D. Grieser, *Exploring Mathematics*, Springer Undergraduate Mathematics Series, https://doi.org/10.1007/978-3-319-90321-7_7

Propositions and predicates

In logic a **proposition** is a statement which can be either true or false. This is called the **truth value** of the proposition.
Here are some examples of propositions:

- A: '$1+1=2$'
- B: 'the square of any even number is even'
- C: '$1=2$'
- D: '5 is negative'
- E: 'Every even number $n > 2$ can be written as the sum of two primes'

The propositions A, B are true, C, D are false; as for E, we don't know whether it is true or false.[1]. 'It is raining' is also a proposition. Look out of the window to see whether it is true.

On the other hand, 'maybe it is raining', 'I hope it will rain soon', 'is 1+1=2?' and 'xy&,%' are not propositions.

The statement 'n is even' is not a proposition because it is not clear what n is. If you plug in a value for n, say $n=3$, then the statement turns into the proposition '3 is even' (which is false). A statement containing one or several variables, which becomes a proposition when we plug in values for the variables, is called a **predicate.** If we denote by $A(n)$ the predicate 'n is even' then the proposition $A(1)$ is false and the proposition $A(2)$ is true, for example.

The variables in a predicate need not correspond to numbers. They can also represent other mathematical objects, for example triangles ('the triangle T has two equal angles') or graphs ('the graph G is connected').

If A is a proposition then the **negation** of A is another proposition, which is sometimes denoted $\neg A$. The negation of '$1+1=2$' is '$1+1\neq 2$'; the negation of 'it is raining' is 'it is not raining'.

[1]Most mathematicians conjecture that it is true. This is called GOLDBACH's conjecture.

A	B	A and B	A or B	$A \Rightarrow B$	$\neg A$
t	t	t	t	t	f
t	f	f	t	f	f
f	t	f	t	t	t
f	f	f	f	t	t

Table 7.1 Truth tables for and, or, if-then and negation. **t** = true, **f** = false

The logical operators 'and', 'or'

From two propositions A, B you can form the new propositions 'A and B', 'A or B'. The truth values of these new propositions are determined by the truth values of A, B as follows:

- 'A and B' is true if and only if *both A and B* are true;
- 'A or B' is true if and only if *at least one* of the propositions A, B is true.

In the examples above 'A and B' is true, 'A and C' is false, 'A or C' is true, and 'C or D' is false. How about 'A or E', 'D or E', 'A and E'?

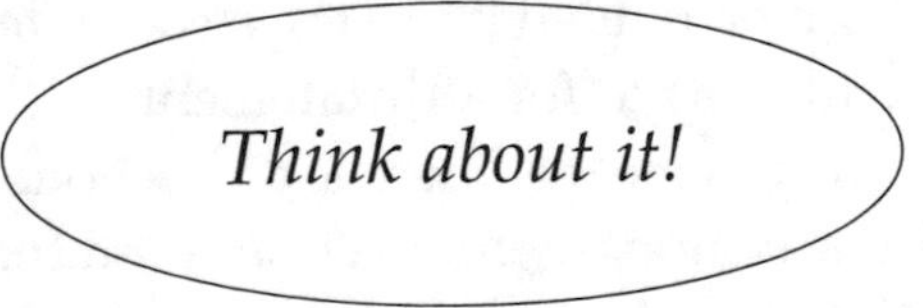

The proposition 'A or E' is true because A is true. Whether the propositions 'D or E', 'A and E' are true we don't know. This depends on the truth value of E.

The proposition 'A or B' in the example is also true: In mathematics 'or' is always inclusive, that is, it is also true if both A and B are true. To express the exclusive 'or' you say 'either A or B' (this would be false for the example). In everyday language it depends on the context whether 'or' is meant in the inclusive or the exclusive sense, and can still be ambiguous.

The logical meaning of 'and', 'or' can be represented in a **truth table**, see the third and fourth column of Table 7.1.

The quantifiers 'for all', 'there is'

Many mathematical statements begin 'For all natural numbers ... ' or 'For all triangles ... ' or 'There is a number for which ... '. In general, suppose $A(n)$ is a predicate and M is a set of values that n can take. Then we can consider the following propositions:

- For all $n \in M$: $A(n)$ holds (in symbols: $\forall n \in M : A(n)$).
- There is an $n \in M$ for which $A(n)$ holds (in symbols: $\exists n \in M : A(n)$).

The expressions 'for all' and 'there is' (or the symbols $\forall$ and $\exists$) are called **quantifiers.** Here are a few examples.

- 'For all $n \in \mathbb{N}$: $n+1 \in \mathbb{N}$' is true.
- 'For all $n \in \mathbb{N}$: $n-1 \in \mathbb{N}$' is false since for $n = 1$ you get $1-1 \in \mathbb{N}$, which is false since 0 is not a natural number[2].
- 'There is a triangle that has two equal angles' is true.
- 'There is a triangle whose angles sum to 200 degrees' is false, since for all triangles the sum of angles is 180 degrees.

To 'refute' a statement means to show that it is false. We see that a 'for all' statement can be refuted by a single counterexample. A 'there is' statement is refuted by a 'for all' statement.

'There is a ... ' always means 'there is at least one ... '. If you want to express that, in addition, there can't be more than one, then you say 'there is precisely one ... ' or 'there is a unique ... '.

Often you use *different wording*, for example: instead of 'for all n: n is positive' we often say 'all $n \in \mathbb{N}$ are positive' or 'for any natural numbers n we have that n is positive' or 'let $n \in \mathbb{N}$; then n is positive'. Instead of 'there is a prime n which is even' we also say 'there is an even prime'. Also, often we say 'there exists' instead of 'there is'.

Important: In mathematics a proposition of the form '$\forall n \in M : A(n)$' is always considered true if M is the empty set. This makes sense since there is nothing that could refute the proposition. In

[2] $\mathbb{N}$ is defined as $\{1,2,3,\dots\}$. Some authors denote the set $\{0,1,2,\dots\}$ by $\mathbb{N}$. In this case the proposition is refuted by $n = 0$. In this book we denote the set $\{0,1,2,\dots\}$ by $\mathbb{N}_0$.

everyday language this is not so: if you say "All of my houses are built of stone" then people will believe that you own at least one house. But logically this does not follow – the statement is true even if you do not own any houses.

You can *change the name of a variable* inside a 'for all' or a 'for is' proposition without changing its truth value: instead of 'for all $n \in \mathbb{N}$: $n+1 \in \mathbb{N}$' you may just as well say 'for all $m \in \mathbb{N}$: $m+1 \in \mathbb{N}$'. Of course you have to change the name at all places referred to by the quantifier. For example, 'for all $m \in \mathbb{N}$: $n+1 \in \mathbb{N}$' is not the same proposition – in fact it is not even a proposition since it is unclear what n is.

This is often useful, even necessary, in arguments where several quantifiers occur. See the proof of the divisibility rules in Section 8.1 for an example.

In many propositions several quantifiers are combined: the everyday wisdom 'every pot has a lid' can be rephrased '*for each* pot *there is* a lid which fits'.

Watch out! Order matters. The proposition 'there is a lid which fits on every pot' would mean something quite different. Formally you could write the two propositions as follows (we usually leave out the colon between quantifiers):

$$\text{'}\forall p \in \{\text{pots}\}\, \exists l \in \{\text{lids}\} : l \text{ fits on } p\text{'}$$
$$\text{'}\exists l \in \{\text{lids}\}\, \forall p \in \{\text{pots}\} : l \text{ fits on } p\text{'}$$

Here is a mathematical example:

'For all $n \in \mathbb{N}$ there is an $m \in \mathbb{N}$ with $m > n$'
'There is an $m \in \mathbb{N}$ such that for all $n \in \mathbb{N}$: $m > n$'

The first proposition is true, the second is false.

The implication: 'if … then …'

From two propositions A, B you can form the new proposition 'if A then B', in symbols '$A \Rightarrow B$'. We also say 'A implies B' and call A the **premise** and B the **conclusion**.[3] The proposition '$A \Rightarrow B$' is false

[3]Sometimes different words are used, for example hypothesis, assumption or antecedent instead of premise, and consequent instead of conclusion.

if A is true and B is false, and is true otherwise. In particular, it is always true if A is false. See the fifth column in Table 7.1. Remember:

> The only way that '$A \Rightarrow B$' can be false is that A is true and B is false.

Why is this reasonable? As an example, consider the proposition:

For all $n \in \mathbb{N}$: if n is divisible by 4 then n is even,

which is clearly true. Then all of the propositions

if 1 is divisible by 4 then 1 is even
if 2 is divisible by 4 then 2 is even
if 3 is divisible by 4 then 3 is even
if 4 is divisible by 4 then 4 is even
etc.

must be true. Let us check the truth values of premise ('... is divisible by 4') and conclusion ('... is even') for each of these propositions: in the first and third line both premise and conclusion are false; in the second line the premise is false and the conclusion is true; and in the fourth line both premise and conclusion are true. These cases correspond to the three lines of the truth table in which '$A \Rightarrow B$' is true. And if we had found a number n which is divisible by 4 but not even, then the implication would have been wrong. This explains all lines in the truth table for '$A \Rightarrow B$'.

Here is another example. A: 'it is raining' and B: 'the street gets wet'. The proposition '$A \Rightarrow B$' says 'if it rains then the street gets wet'. This could also be phrased as '*every time* it rains, the street gets wet'.[4] This is only wrong if there is a time when it rains but the street remains dry. If it never rains then the statement is still true, no matter whether the street gets wet or not – since nothing is claimed in that case.

As in these examples, propositions of the form 'if ... then ...' are almost always part of a 'for all' clause, even if this is not stated explicitly. Therefore, whenever you see 'if ... then ...' you should add something like 'every time ...'.

[4] Very precisely: Let $A(t), B(t)$ be the propositions, for a point in time t: 'it is raining at time t' and 'the street is getting wet at time t'. Then '$A \Rightarrow B$' is short for the 'for all' proposition '$\forall t : A(t) \Rightarrow B(t)$'.

We can rephrase implications in various ways: instead of saying 'if it rains then the street gets wet' we can say 'if the street remains dry then it is not raining', or 'it does not happen that it rains and the street remains dry'. Formally this says that for any two propositions A, B the following three propositions are equivalent:

$$A \Rightarrow B \qquad \neg B \Rightarrow \neg A \qquad \neg(A \text{ and } \neg B).$$

This is the basis of **indirect proof** and of **proof by contradiction**, see below. Here, 'equivalent' means that for any truth values of A, B the three propositions above have the same truth values. You can check this directly using Table 7.1. We also say that they are **logically equivalent.**

But note that $A \Rightarrow B$ is *not* logically equivalent to $\neg A \Rightarrow \neg B$. In the example this would be 'if it is not raining then the street remains dry' – this is clearly a different proposition (which is false since the street could also get wet from a lawn sprinkler). Convince yourself using the truth table. Similarly, $A \Rightarrow B$ is not logically equivalent to $B \Rightarrow A$.

You hear this sort of fallacy quite often in everyday life, for example like this: 'If it rains then the street gets wet. It's not raining. Therefore the street remains dry.' or 'If it rains then the street gets wet. The street is wet. Therefore it must be raining.' Both conclusions are logically incorrect.

Necessary and sufficient

Instead of '$\forall n \in M : A(n) \Rightarrow B(n)$' we also say '$A(n)$ is sufficient for $B(n)$, for $n \in M$' and '$B(n)$ is necessary for $A(n)$, for $n \in M$'. Often we leave out 'for $n \in M$' when M is clear from the context.
These phrases emphasize different points of view:

- $A(n)$ is sufficient for $B(n)$: we are looking for conditions on n which ensure that $B(n)$ is true; if we have proved '$\forall n \in M : A(n) \Rightarrow B(n)$' then we know that $A(n)$ is such a condition.

- $B(n)$ is necessary for $A(n)$: we are looking for conditions on n which must necessarily hold for $A(n)$ to be true; if we have proved '$\forall n \in M : A(n) \Rightarrow B(n)$' then we know that $B(n)$ is such a condition.

As an example consider the propositions $A(n)$: 'the decimal representation of n ends in a zero' and $B(n)$: 'n is even' about natural numbers $n \in \mathbb{N}$. The true proposition '$\forall n \in \mathbb{N} : A(n) \Rightarrow B(n)$' can be restated as follows:

- $A(n)$ is sufficient for $B(n)$: in order to know that n is even, it suffices to know that n ends in a zero.
- $B(n)$ is necessary for $A(n)$: if n ends in a zero then it must necessarily be even.

But $B(n)$ is not sufficient for $A(n)$: an even number need not end in a zero. Similarly, $A(n)$ is not necessary for $B(n)$. Pay attention to the reversal of order when using 'necessary'.

We can summarize the meaning of 'necessary' and 'sufficient' as follows: If $A(n), B(n), C(n)$ are predicates then $A(n) \Rightarrow B(n) \Rightarrow C(n)$ means that $A(n)$ is sufficient for $B(n)$ and that $C(n)$ is necessary for $B(n)$.

We usually use these words when we are investigating a complicated property of certain objects, and trying to find a condition which is easy to check and which is logically related to this property.

Here is an example: Consider the following proposition $A(G)$ about a graph G: 'G can be drawn in the plane without edge intersections'. The solution of Problem 4.6 shows that a necessary condition for $A(G)$ is: G has no 5 vertices which are all joined by edges. However, this condition is not sufficient: suppose G contains 6 vertices $E_1, E_2, E_3, F_1, F_2, F_3$ where each E_i is joined to each F_j by an edge. Then G does not satisfy $A(G)$ either, see Exercise E 4.6.[5]

Equivalence: 'if and only if'

We call propositions A, B (logically) **equivalent** and write $A \Leftrightarrow B$ if both $A \Rightarrow B$ and $B \Rightarrow A$ are true. We also say 'A holds **if and only if** B holds' or 'A is **necessary and sufficient** for B'.

As for the implication, this usually occurs inside a 'for all' clause: $\forall n \in M : A(n) \Leftrightarrow B(n)$.

[5]Amazingly, in a sense this is all that can happen: using these two special cases one can formulate a condition which is *necessary and sufficient* for $A(G)$. This is KURATOWSKI's theorem, see (Aigner, 1987) for example.

Negation of composite propositions

When negating a composite proposition you have to be very careful. It is worth clarifying this in simple examples and then memorising how it works.

In order to find the negation of a proposition A you ask yourself: what precisely must be true in order for A to be false? What do I need to show in order to refute A?

This leads to the **inversion rule** for negations:

> When negating, interchange 'and' with 'or' and $\forall$ with $\exists$, and pull the negation sign to the end.

This means:[6]

the proposition	is equivalent to the proposition
$\neg(A \text{ and } B)$	$\neg A \text{ or } \neg B$
$\neg(A \text{ or } B)$	$\neg A \text{ and } \neg B$
$\neg(\exists n \in M : A(n))$	$\forall n \in M : \neg A(n)$
$\neg(\forall n \in M : A(n))$	$\exists n \in M : \neg A(n)$

Examples

- The proposition 'yesterday it rained *and* snowed' is false if and only if it did not rain *or* it did not snow.
- The proposition 'yesterday it rained *or* snowed' is false if and only if it did not rain *and* it did not snow.
- The proposition 'it rains *every* day' is false if and only if *there is* a day when it does not rain.
- The proposition 'this week *there was* a day when it rained' is false if and only if it was dry *every* day of the week.

The inversion rule is especially useful when negating more complex combinations of propositions. You apply it several times:

The negation of 'for all $n \in \mathbb{N}$ there is an $m \in \mathbb{N}$ with $m > n$'
is 'there is an $n \in \mathbb{N}$ so that for all $m \in \mathbb{N}$: $m \leq n$'.

[6] '$\neg A$ or $\neg B$' is to be understood as '$(\neg A)$ or $(\neg B)$'

Be sure you understand that this is the logically correct negation. Here is an another example: the negation of 'every pot has its lid' is 'there is a pot which does not have a lid'. More systematically: the negation of 'for every pot there is a lid that fits' is 'there is a pot for which all lids don't fit'.

You can also check this purely formally by pulling the $\neg$ sign to the right step by step, using the inversion rule in each step:

$$\begin{aligned}
\neg(\forall n \in \mathbb{N} \quad \exists m \in \mathbb{N}: \quad m > n) &\iff \\
\exists n \in \mathbb{N}\ \neg(\exists m \in \mathbb{N}: \quad m > n) &\iff \\
\exists n \in \mathbb{N} \quad \forall m \in \mathbb{N}: \neg(m > n) &\iff \\
\exists n \in \mathbb{N} \quad \forall m \in \mathbb{N}: \quad m \leq n &
\end{aligned}$$

In newspapers and politicians' speeches you will find many incorrect negations. For example, the statement 'For all t: $A(t)$' is often negated as 'For all t: $\neg A(t)$', but this means something quite different.

7.2 Proofs

A proof is a logically complete justification of a proposition. As long as we have not proven a proposition it is possible that it is false – even if it is supported by many examples.

Example

In the year 1637 FERMAT investigated the numbers $F_n = 2^{(2^n)} + 1$. He noticed that F_0, F_1, F_2, F_3 and F_4 are prime numbers. For example: $F_0 = 2^{(2^0)} + 1 = 2^1 + 1 = 3$, $F_1 = 2^{(2^1)} + 1 = 2^2 + 1 = 4 + 1 = 5$, $F_2 = 2^{(2^2)} + 1 = 2^4 + 1 = 16 + 1 = 17$. Therefore he conjectured that F_n is prime for all n.

It was almost 100 years until EULER discovered, in 1732, that $F_5 = 4294967297$ is not prime: it is divisible by 641. Today the numbers F_n are called FERMAT numbers.

This shows that examples can be quite misleading.[7] So we need a proof if we want to be sure that a proposition is true.

[7] See the Exercises E 1.10 and E 5.24 for other such problems.

In addition, a proof can give you a deeper understanding of a fact. A good example is the second/third proof of the formula $|\mathcal{P}(\{1,\ldots,n\})| = 2^n$ in Section 5.2.

General types of proofs

Often we want to prove propositions of the form $A \Rightarrow B$.[8][9] For this we can proceed in different ways.

Direct proof The proof is composed of steps whose validity is obvious or which have been proved before: instead of proving '$A \Rightarrow B$' directly, we prove '$A \Rightarrow A_1 \Rightarrow A_2 \Rightarrow \ldots \Rightarrow B$'.

Indirect proof We suppose that the conclusion B is false and deduce from this that the premise A must also be false. So instead of '$A \Rightarrow B$' we prove '$\neg B \Rightarrow \neg A$', which is logically equivalent. Of course the latter proof can be composed of several steps.

Proof by contradiction A proof by contradiction is very similar to an indirect proof. We suppose that A is true and B is false and derive a contradiction (a false statement) from this.

So we prove '(A and $\neg B$) $\Rightarrow$ **f**'. Since true propositions can imply only true propositions, it follows that 'A and $\neg B$' is false. This is logically equivalent to '$A \Rightarrow B$'.

Sometimes there is no premise: We want to prove a proposition B. Then a direct proof concludes B from known propositions, and a proof by contradiction derives a false statement from $\neg B$: $\neg B \Rightarrow$ **f**.[10]

Here are two more general types of proofs.

Refutation by counterexample In order to refute a 'for all' proposition (i.e. in order to prove that it is false) it suffices to give one counterexample.

[8]Instead of 'to prove' we also say 'to show'.

[9]More precisely, these are propositions of the form 'for all X: $A(X) \Rightarrow B(X)$', for certain mathematical objects X (for example, natural numbers or graphs). For better readability we use the abbreviated form $A \Rightarrow B$ in the sequel. Compare the discussion on implication in the previous section.

[10]Strictly speaking, there is no indirect proof in this case. However, the proof by contradiction is sometimes called indirect proof.

Mathematical induction This can often be used to prove propositions of the form 'For all $n \in \mathbb{N} \ldots$'. See Chapter 3.

We have already seen a refutation by counterexample: the proposition 'for all $n \in \mathbb{N}_0$: F_n is a prime number' about the FERMAT numbers was disproved by showing that F_5 is not a prime.

Remark

In a proof you can use propositions which have been proved before or which are considered to be known in a given context. In a strictly axiomatic build-up of mathematics, which you learn when you study mathematics at university, you are only allowed to use axioms, or propositions which have been proved from the axioms beforehand.[11] This assures you that all proven propositions are true – assuming only that the axioms are true. You have to begin somewhere, and proceeding in this way allows you to make this 'base' very small.

Here is a simple example of a direct proof.

Theorem The sum of two even numbers is even.

We first reformulate this to show clearly the premise and the conclusion:

Claim:

Let n, m be even numbers. (premise)
Then $n + m$ is even. (conclusion)

Proof.
Since n, m are even, there are $j, k \in \mathbb{Z}$ so that $n = 2j$ and $m = 2k$ (this is the definition of 'even'). Then $n + m = 2j + 2k = 2(j + k)$. We have $j + k \in \mathbb{Z}$. Therefore, $n + m$ is also of the form $n + m = 2q$ with $q \in \mathbb{Z}$ and hence an even number. q. e. d.

[11] An **axiom** is a proposition which is postulated to be true in a given context. For example, one of the axioms of plane geometry is the **parallel postulate:** Given any line g and any point P not lying on g there is a unique line g' which contains P and does not intersect g. As foundation for a mathematical theory one chooses as few axioms as possible.

Proof analysis: This was a **direct proof.** We first applied the definition of 'even'. Then there was one intermediate step: the calculation of $n + m$ using the distributive law. Then we applied the definition again.

We now consider some common types of propositions and the proof patterns which are typical for them.

Proofs of formulas

Formulas express general relations between numbers or other mathematical objects. In previous chapters you have already encountered many formulas, for example $2^n = \sum_{k=0}^{n} \binom{n}{k}$ or counting formulas.

Sometimes you conjecture a formula when investigating a problem (for example in Problem 1.3). Then you need to prove it in general.

Typical proof patterns for formulas:

- Direct derivation by combining known formulas
- Mathematical induction (see Chapter 3)
- Counting in two ways (see Chapter 5)

When penetrating deeper into mathematics you will get to know many more ways to derive or prove formulas. One of them is the solution of the FIBONACCI recursion in Chapter 2.

The point of view of modern mathematics is that formulas are usually expressions of deep relationships between structures (for example geometric objects, algebraic structures etc.). An example of this is counting in two ways.

Proofs of existence

Many mathematical questions are about the existence of certain objects: does an equation have a solution? Are there numbers, triangles, graphs, ... having certain properties?

As an example, we ask whether the following equations have a solution (in real numbers):

a) $x^5 + x = 2$: we can exhibit a solution: $x = 1$.

b) $x^2 + 1 = 0$: this equation does not have a solution since the left-hand side is positive for all real numbers x.

c) $x^5 + x = 1$: this is harder. Try out a few values for x. It is very unlikely that you will find a solution. But the equation does have a solution. However, this can only be proved using the intermediate value theorem from analysis.[12]

So the simplest way to prove existence is to **exhibit** the desired object. But often we want to prove a proposition of the kind 'for all $n \in M$ there is ...' where M is an *infinite* set. That is, we need infinitely many proofs of existence – or one which works for all n. The simplest way to do this is to give a **general construction.**

Here are two examples:

Theorem For every odd number $n \in \mathbb{N}$ there are two square numbers whose difference equals n.

Try to find a general construction, i.e. a prescription for finding such a pair of squares from n. Use problem solving strategies (for example, looking at small examples).

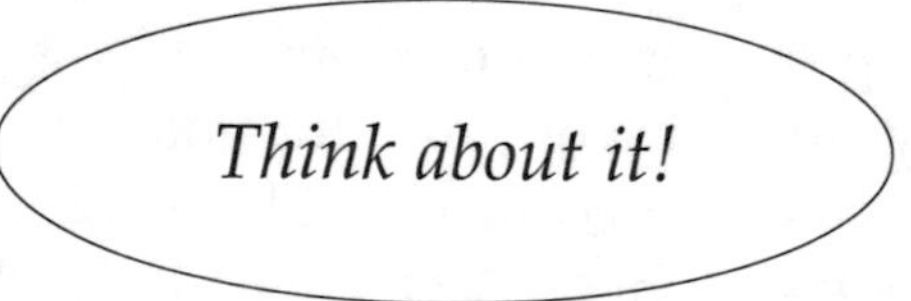

Proof.
Since n is odd we can write it as $n = 2m + 1$ where $m \in \mathbb{N}_0$. Then $n = (m+1)^2 - m^2$.
q. e. d.

Proof analysis: We needed to supply a proof of existence for each of the infinitely many odd numbers n. This was accomplished by a general construction, in this case a general formula for the two squares.

[12] Idea: for $x = 0$ we have $x^5 + x = 0 < 1$ and for $x = 1$ we have $x^5 + x = 2 > 1$, therefore there must be an x between 0 and 1 satisfying $x^5 + x = 1$. The intermediate value theorem says that the last conclusion is correct, based on the fact that the function $f(x) = x^5 + x$ is continuous.

Theorem For any triangle there is a point that is the same distance from all three vertices.

Investigation

How could we find such a point? Let us first simplify the problem: instead of requiring the same distance from three points we content ourselves with the same distance from *two* points. So we ask: Given two points A, B in the plane, which points are the same distance from A and from B? Maybe you remember the answer from school: these are precisely the points on the perpendicular bisector of the line segment $\overline{AB}$.

How do we get from same distance from two points to same distance from *three* points A, B, C?

Think about it!

By first considering the points A, B and then the points B, C: the point we are looking for must lie on the perpendicular bisector of $\overline{AB}$ and on the perpendicular bisector of $\overline{BC}$, so it must be the intersection point of these two lines. Do they intersect? Yes, if they are not parallel. Why are they not parallel? Since $\overline{AB}$, $\overline{BC}$ are not parallel if ABC is a triangle.

We write up the proof in order.

Proof.
Denote the vertices of the triangle by A, B, C. Let g be the perpendicular bisector of the side $\overline{AB}$ and h the perpendicular bisector of the side $\overline{BC}$. Every point of g is the same distance from A and B, and every point of h is the same distance from B and C.

Since the sides $\overline{AB}$, $\overline{BC}$ are not parallel, the lines g and h are not parallel, therefore they have a point of intersection. We call this point P.

Since P lies on g it is the same distance from A and B. Since P lies on h it is the same distance to B and C. Therefore P is the same distance from A, B and C. q. e. d.

Proof analysis: We wanted to show the existence of a special point for any triangle. For this we made a general construction. For finding the construction it was useful to first consider a simpler variant of the problem.

Sometimes we want to prove the existence not only of a *single* object of a certain type, but of *infinitely many* such objects. This turns out to be similar to proving a proposition of the type 'for all $n \in M$ there is ...'. Here is a famous example.

Theorem There are infinitely many prime numbers.

Proof.
We give a procedure which does the following: for an arbitrary finite set of primes it produces a prime which is not contained in this set.

Let $p_1, p_2, \ldots, p_m$ be primes. Consider the number $n = p_1 \cdot p_2 \cdots p_m + 1$. This is clearly bigger than each of the numbers $p_1, \ldots, p_m$. So if n is prime then we are done.

If n is not prime then it is divisible by a prime.[13] Call this prime p. Since n is divisible by p, the number $n - 1$ is not divisible by p. Therefore $p_1 \cdot p_2 \cdots p_m = n - 1$ is not divisible by p, so p cannot be one of the primes $p_1, \ldots, p_m$.

We have proved that for any finite set of primes there is another prime not contained in the set. Therefore there are infinitely many primes.

q. e. d.

Proof analysis: This was a constructive proof of existence. We could have formulated the same idea as proof by contradiction: suppose there were only finitely many primes $p_1, \ldots, p_m$. Then construct a prime as above, which does not appear among them. This contradicts the assumption that $p_1, \ldots, p_m$ are all the primes.[14]

[13]Why? Let p be the smallest integer which divides n and which is bigger than 1. Then p must be a prime number, since otherwise we could write $p = ab$ with $1 < a < p$, and a would divide n and would be smaller than p, contradicting the choice of p.

[14]You will find this indirect argument in many books. However, it is better to phrase arguments in a direct way whenever possible. This makes them easier to understand. Principle: use as few negations as possible.

The limits of general constructions: In many problems of the type "Show that for all $n \in \mathbb{N}$ there is an X satisfying ..." (where X is some sort of mathematical object) you will experience the following: You can prove the claim for many special cases (say $n = 1, 2, 3, 4, 5$) by finding an X. But you don't find a general construction which works for all n, and you don't find a pattern that could lead you to such a construction. You will encounter such problems in the following chapters, and you will get to know two far-reaching ideas for such situations: the pigeonhole principle and the extremal principle. Let us summarize:

Typical patterns for proofs of existence:

The simplest way to prove existence of an object is to exhibit it. In order to prove a proposition of the type 'for all $n \in M$ there is ...' you can give a general construction which works for all n, or try one of the following strategies:

- The pigeonhole principle (see Chapter 9)
- The extremal principle (see Chapter 10)

Of course general proof strategies like proof by contradiction or mathematical induction (here in the form of an inductive construction) can be useful as well.

In each area of mathematics you will find more specific tools for proofs of existence. For example, in analysis the intermediate value theorem and fixed point theorems, and in linear algebra (and other areas) dimensional arguments (a homogeneous linear system of equations with more variables than equations must have a non-trivial solution).

We have already encountered a special type of proof of existence: refutation by counterexample. This is a case of proof of existence by exhibiting an object: the negation of 'for all $n \in \mathbb{N}$: $A(n)$' is 'there is an $n \in \mathbb{N}$ for which $A(n)$ is false', and this can be proved by exhibiting an n for which $A(n)$ is false.

When disproving FERMAT's conjecture that all FERMAT numbers are prime EULER used the principle 'proof of existence by exhibiting' twice: he exhibited the FERMAT number F_5 which is not prime; and in order to prove this he exhibited the non-trivial factor 641 of F_5. So the second proof of existence is nested inside the first.

Proofs of nonexistence and impossibility

To prove that something does not exist or is impossible is a fascinating kind of problem and can be quite difficult. Here are some examples of such statements:[15]

- It is impossible to join five points in the plane pairwise in such a way that the connecting lines don't intersect.
- A power of 2 cannot be represented as a trapezoidal number.
- There is no algorithm for calculating the prime factorisation of a large integer in acceptable time.
- For all natural numbers $n > 2$ the equation $x^n + y^n = z^n$ has no solution with $x, y, z \in \mathbb{N}$.

The first two examples are Problem 4.6 and Problem 6.3. The third example needs to be made precise, for example like this: there is no algorithm which calculates the prime factorisation[16] of an arbitrary 1000 digit number (in the decimal system) in less than 1000 years, on current computer hardware. This could be read as a statement about known algorithms, and read this way it is true. However, whether such an algorithm *cannot exist* at all is a difficult, unsolved problem: So far nobody has found such an algorithm, but no one has been able to prove it cannot exist. See the end of Section 11.4 for the practical relevance of this problem.

The fourth example is the famous FERMAT problem. As early as 1637 FERMAT claimed that there cannot be a solution, but this was proved only in 1995.

Two other famous examples are the impossibility of trisecting an arbitrary angle using only ruler and compass, and GÖDEL's incompleteness theorem, which says that it is impossible to prove mathematically that mathematics is free of contradictions. See Wikipedia for explanations and references.

[15]Nonexistence and impossibility are really the same thing. For example, the second statement can be rephrased as 'there does not exist a representation of a power of 2 as ...' or as 'it is impossible to represent a power of 2 as ...'.

[16]We will discuss prime factorisation in detail in Chapter 8.

Here is a classical nonexistence statement.

Theorem The number $\sqrt{2}$ is irrational.

Investigation

Let us first see that this is a statement of nonexistence: the claim is that $\sqrt{2}$ is not rational, that is, that there are no natural numbers p, q satisfying $\sqrt{2} = \frac{p}{q}$.

How could we prove that? We could try a proof by contradiction: wesuppose there were such p, q, then we try to derive more and more consequences from this until we arrive at a contradiction.

So assume there are $p, q \in \mathbb{N}$ with $\sqrt{2} = \frac{p}{q}$. What could we do with this equation? Can we simplify it?

Think about it!

There are two complications: the fraction and the square root. So let's multiply by q to get rid of the fraction and then square to get rid of the root:

$$\sqrt{2} = \frac{p}{q} \Rightarrow \sqrt{2}\,q = p \Rightarrow 2q^2 = p^2 .$$

How could we continue, what can we conclude from this?

The equation $2q^2 = p^2$ implies that p^2 is even. Then p must also be even (see below for a complete proof). We make this explicit by writing $p = 2r$ for some natural number r.

We square and get $p^2 = 4r^2$. We plug this into $2q^2 = p^2$ and obtain $2q^2 = 4r^2$. Now we can divide by 2 and get

$$q^2 = 2r^2 .$$

By the same argument as before q must be even. So both p, q must be even, so we can cancel the factor 2 in the original fraction $\frac{p}{q}$.

But what if we had started with a fraction $\sqrt{2} = \frac{p}{q}$ in lowest terms? We could certainly choose p, q to do this. Then we would have the desired contradiction.

Let us write up the proof in order. First we prove a lemma.[17]

Lemma Let $n \in \mathbb{N}$. If n^2 is even then n is even.

Proof.
We prove this indirectly. Suppose n was odd. Then we could write $n = 2m + 1$ where $m \in \mathbb{N}_0$. Then we would get $n^2 = (2m+1)^2 = 4m^2 + 4m + 1 = 2(2m^2 + 2m) + 1$, so n^2 would be odd. q. e. d.

We now prove that $\sqrt{2}$ is irrational.

Proof.
Suppose $\sqrt{2} = \frac{p}{q}$ with $p, q \in \mathbb{N}$. We may assume that the fraction is reduced to lowest terms, so that p and q have no common factor. We multiply by q and square to obtain $2q^2 = p^2$. Therefore p^2 is even, so p is even by the lemma. Write $p = 2r$ with $r \in \mathbb{N}$. Squaring this and plugging in we obtain $2q^2 = 4r^2$, hence $q^2 = 2r^2$. Therefore q^2 is even, hence q is even by the lemma. Therefore p and q are both even, so they have the common factor 2. This is a contradiction to the assumption that the fraction $\frac{p}{q}$ was in lowest terms. Therefore $\sqrt{2}$ cannot be rational. q. e. d.

Proof analysis: This was a proof by contradiction.

Typical patterns for proofs of nonexistence/impossibility:

- Proof by contradiction
- The invariance principle (see Chapter 11)

Exercises

1 E 7.1 Take a newspaper and find at least three examples of logically incorrect reasoning.

1-2 E 7.2 Each of the five cards in Figure 7.1 has a letter on one side and a number on the other side. How many cards do you need to

[17] A lemma (from the Greek) is an auxiliary proposition which is used in the proof of a theorem.

turn over in order to find out whether the following proposition is true: *If the card has an even number on one side then the other side shows a vowel.*

Figure 7.1 Exercise E 7.2

E 7.3 Consider a colouring of the natural numbers using the colours red and blue. Formally, this is a map $f : \mathbb{N} \to \{\text{red}, \text{blue}\}$. Also consider the following proposition: 1-2

For each blue number there is a larger number which is red.

Which of the following propositions follow from this?

a) There is a blue natural number.

b) $\exists n \in \mathbb{N} : f(n) = \text{red}$

c) For each red number there is a smaller number which is blue.

d) For each red number there is larger number which is red.

e) $|\{m \in \mathbb{N} : f(m) = \text{red}\}| = \infty$

f) If $f(1) = \text{blue}$ then there is $n \in \mathbb{N}$ such that $f(n) = \text{blue}$ and $f(n+1) = \text{red}$.

E 7.4 Suppose a connected figure consisting of unit squares is given. Decide for each of the following propositions which of the other propositions it implies. Give proofs. 1-2

a) The number of squares is three.

b) The number of squares is divisible by three.

c) The figure can be tiled by tiles of the form □□□.

d) The figure can be tiled by tiles of the form ▟.

E 7.5 A restaurant is good if each guest finds a least three dishes on the menu that he likes. When is a restaurant not good? 1

2 **E 7.6** Formulate the negations. (For b) and c) let $S_1 \subset \mathbb{N}$ and $S_2 \subset \mathbb{N}$.)

a) All trees in this garden bear at least one apple.

b) For each $n \in S_1$ there is $m \in S_1$ such that $m > n$.

c) For each $n \in S_2$: if n is divisible by three then n is odd.

Construct sets $S_1 \subset \mathbb{N}$ and $S_2 \subset \mathbb{N}$ so that the propositions b) and c) are true. Formulate a shorter proposition equivalent to b).

2 **E 7.7** For natural numbers m, n we write $m|n$ (in words: m divides n) if n is divisible by m. Formulate the following proposition in as simple words as possible.

$$\forall n \in \mathbb{N} : (\exists m \in \mathbb{N} : 1 < m < n \text{ and } m|n) \Rightarrow (\exists m \in \mathbb{N} : 1 < m^2 \leq n \text{ and } m|n)$$

Is the proposition true? Formulate its negation formally and in words.

1-2 **E 7.8** Formulate the proposition 'Each person has a secret' using quantifiers. Here a secret is a fact that no other person knows about.

1-2 **E 7.9** Let $x, y \in \mathbb{N}$. Which of the following equivalences hold?

$$xy \text{ odd} \overset{?}{\Leftrightarrow} x \text{ and } y \text{ odd}$$

$$xy \text{ even} \overset{?}{\Leftrightarrow} x \text{ and } y \text{ even}$$

$$xy \text{ even} \overset{?}{\Leftrightarrow} x \text{ or } y \text{ even}$$

2 **E 7.10** Decide for each proposition on the right whether it is necessary and/or sufficient for the proposition to the left of it.

Here a, b are always real numbers.

x is a man.	x is a mammal.
$a < b$	$\exists c \in \mathbb{R} : a < c$ and $c < b$
$a < b$	$\exists c \in \mathbb{N} : a < c$ and $c < b$
A figure consists of an even number of identical squares.	The figure can be tiled by dominoes.

E 7.11 Which of the following propositions about a triangle T are necessary, which are sufficient and which are both necessary and sufficient for the truth of the proposition 'D is equilateral'? 2

a) 'T has two equal angles'

b) 'All sidelengths of T are equal to 1'

c) 'All angles of T are at least 60 degrees.'

E 7.12 Let a, b be rational numbers satisfying $a < b$. Prove that there is a rational number r such that $a < r < b$. Give a direct proof by construction. (See also Exercise E 10.7 for the case where a, b are real numbers.) 1-2

E 7.13 Prove that among any seven numbers in $\{1, 2, \ldots, 128\}$ there are x, y with $x < y \leq 2x$. 1-2

E 7.14 Find a necessary and sufficient condition on a natural number v for there to be a graph having v vertices all of whose degrees are odd. 2

Prove that your condition is necessary and sufficient.

E 7.15 Because $\sqrt{2} - 1 = 0.41\ldots$ is less than 1, the powers $(\sqrt{2} - 1)^n$ approach zero when n gets large. Use this and the general binomial theorem (see Exercise E 5.15) to give another proof that $\sqrt{2}$ is irrational. 2

Also prove that $\sqrt{m}$ is irrational whenever m is a natural number which is not a square.

8 Elementary number theory

Number theory is the mathematics of the integers: it is about divisibility, prime numbers etc. Integers are very concrete objects that you have played with since you were a small child. Therefore number theory is very suitable for your exploration of mathematics, and you will find number-theoretic problems in many places in this book. Number theory has many faces: some of the hardest problems of mathematics, still unsolved today, are stated in simple number-theoretic terms.

In this chapter we collect some of the basic definitions and theorems of number theory. Most of the theorems will seem quite natural from a few examples, and turning your number intuition into rigorous proofs will be good exercise for you. You will also learn about the notion of congruence, which magically turns some difficult problems into easy exercises.

8.1 Divisibility, prime numbers and remainders

The most important concept of number theory is divisibility. What does divisibility mean? For example, 15 is divisible by 3 because dividing 15 by 3 leaves no remainder: $15 = 5 \cdot 3$. This leads to the following definition.

Definition Let $a, n \in \mathbb{Z}$. We write $n|a$ if there is a number $q \in \mathbb{Z}$ satisfying $a = qn$.

Some ways to express this are: n **divides** a, or n **is a divisor of** a, or a **is a multiple of** n. For example, $1|a$ for all $a \in \mathbb{Z}$, and $2|a$ means that a is even.

Note that negative numbers and zero are permitted. For example, the divisors of -5 are the numbers $1, -1, 5, -5$. Zero is divisible by every integer n since $0 \cdot n = 0$. Here are the most important rules for divisibility:

D. Grieser, *Exploring Mathematics*, Springer Undergraduate Mathematics Series, https://doi.org/10.1007/978-3-319-90321-7_8

Lemma **(Rules for divisibility)** Let $a, b, n \in \mathbb{Z}$.

a) If $n|a$ and $a|b$ then $n|b$.

b) If $n|a$ and $n|b$ then $n|a+b$ and $n|a-b$.

Read $n|a+b$ as $n|(a+b)$, and similarly for $n|a-b$.

Proof.
This follows directly from the definition. Let us prove a):
$n|a$ means that there is $q \in \mathbb{Z}$ with $a = qn$.
$a|b$ means that there is $q' \in \mathbb{Z}$ with $b = q'a$.
Plugging the first equation into the second we get $b = q'(qn) = (q'q)n$. Because $q'q$ is an integer, this implies $n|b$. The proof of b) is similar (exercise).

q. e. d.

Proof analysis: This was a **direct proof.** Why did we write q' in the second line of the proof, not q as in the definition? Because q was already in use.

Definition A number $p \in \mathbb{N}$ is called a **prime number** (or a **prime**) if it has precisely two positive divisors.

For example, 2 has the positive divisors 1 and 2, hence is prime, but 4 has the positive divisors 1, 2 and 4, hence is not prime. 1 has only one positive divisor (namely 1), hence is not prime. To phrase the definition in this way (which excludes 1 from being prime) is only a matter of convention. We will see below that it is a useful convention, because only then is the prime factorisation unique.
Prime numbers have the following important property.

Lemma **(Euclid's lemma)** Let $a, b \in \mathbb{N}$ and p a prime number. If p divides the product ab then p divides a or b:

$$p|ab \Rightarrow p|a \text{ or } p|b$$

This is familiar for $p = 2$: If ab is even then a or b must be even. The proof for general p is trickier than you might expect, see Exercise E 8.12. Be careful: If p is not a prime then the statement is wrong. For example, $4 | 2 \cdot 6$ but $4 \nmid 2$ and $4 \nmid 6$. Here $n \nmid a$ means that n does *not* divide a.

The lemma implies: If p is a prime and if p divides a product $a_1 \cdots a_l$ then p divides one of the factors a_i (proof by induction as exercise).

The following theorem is so important that it is called the **fundamental theorem of arithmetic.**

Theorem **(Prime factorisation)** Every natural number bigger than one has a prime factorisation. It is unique up to the order of the factors.
This means:

Let $n \in \mathbb{N}$, $n > 1$.

(i) There are $m \in \mathbb{N}$ and prime numbers $p_1, \ldots, p_m$ with $n = p_1 \cdots p_m$.

(ii) If, in addition, $q_1, \ldots, q_l$ are prime numbers satisfying $n = q_1 \cdots q_l$ then $l = m$ and the numbers $q_1, \ldots, q_m$ are the same as the numbers $p_1, \ldots, p_m$, up to reordering.

The same prime can occur multiple times. Then it appears as many times among the p_i as among the q_j.
Examples: $2 = 2$, $10 = 2 \cdot 5 = 5 \cdot 2$, $999 = 3 \cdot 3 \cdot 3 \cdot 37$.

We have used the prime factorisation already in the solutions of Problems 1.2 and 6.3. Only the following proof makes these solutions complete.

Proof.
Since this is a statement about all natural numbers $n > 1$ it is natural to look for an inductive proof. Try it yourself first!

Think about it!

Proof of (i) by induction on n:
Base case: $n = 2$ is a prime number, so we can take $m = 1$, $p_1 = 2$.
Inductive hypothesis: Let $n \in \mathbb{N}$, $n > 2$. Assume that (i) holds for all natural numbers bigger than one and less than n.
Inductive step: If n is prime then we are done. If n is not a prime then we can write $n = ab$ where $1 < a, b < n$. Applying the inductive hypothesis we know that a, b are primes or products of primes. Then $n = ab$ is also a product of primes.

Proof of (ii) by induction on n:
Base case: The claim is obvious for $n = 2$.
Inductive hypothesis: Let $n \in \mathbb{N}$, $n > 2$. Assume that (ii) holds for all natural numbers bigger than one and less than n.
Inductive step: Let $n = p_1 \cdots p_m = q_1 \cdots q_l$. Then p_1 divides the product $q_1 \cdots q_l$, so by EUCLID's lemma (or rather the remark after it) p_1 divides one of the factors q_i. By reordering these factors we may arrange that $p_1 | q_1$. Because q_1 is a prime, this implies $p_1 = q_1$. Then we can divide by p_1 and obtain $p_2 \cdots p_m = q_2 \dots q_l$. This number is smaller than n, so we can apply the inductive hypothesis. It says that $m - 1 = l - 1$, hence $m = l$, and that $p_2, \dots, p_m$ agree with $q_2, \dots, q_m$ up to reordering. The claim follows. q. e. d.

We now discuss division with remainder.[1] You know this from school. For example: 20 divided by 3 equals 6 with a remainder of 2. This means $20 = 6 \cdot 3 + 2$. Here it is essential that the remainder (2) is smaller than the divisor (3). In general, it works like this:

Theorem **(Division with remainder)** Let $a \in \mathbb{Z}$, $n \in \mathbb{N}$. Then there are unique numbers $q, r \in \mathbb{Z}$ satisfying

$$a = qn + r \quad \text{and} \quad 0 \leq r < n\,.$$

Here q is called the **quotient** and r the **remainder.**
We also call r the remainder of a **modulo** n.

In words: a divided by n equals q with a remainder of r. The possible remainders when dividing by n are

$$0, 1, \dots, n - 1.$$

[1] This is also known as Euclidean division.

Proof.
Initial ideas: How do you think about division with remainder? Maybe like this: You take as many n as you can fit into a; there are q of them, and what remains is the remainder r. The remainder is less than n because otherwise another n would fit into a. Therefore q and r exist. They are unique, for if we used another q then the remainder would be negative or bigger than n.

Now we just need to express this in mathematical notation. Try it yourself!

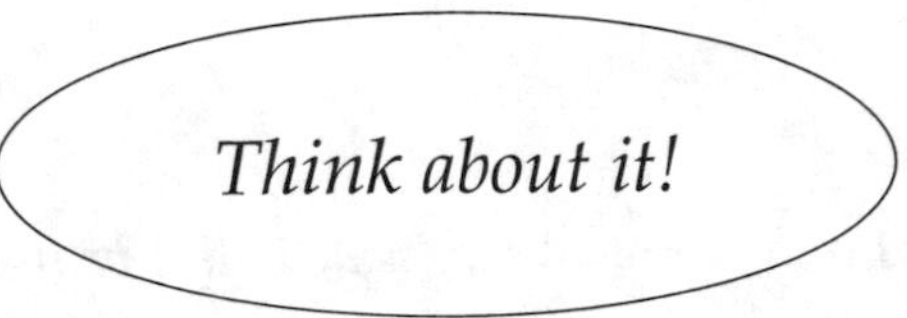

We first prove the existence of q and r. Let q be the largest integer satisfying $qn \leq a$. Let $r = a - qn$. Then $qn \leq a$ implies $r \geq 0$. Also $r < n$ since otherwise $r \geq n$, hence $a - (q+1)n = a - qn - n = r - n \geq 0$ and thus $(q+1)n \leq a$, which would contradict the choice of q as *largest* integer satisfying $qn \leq a$.

Now we prove the uniqueness of q and r. Let q and r be constructed as above, and suppose that $a = q'n + r'$ is any representation. If $q' > q$ then $q'n > a$ because of the maximality of q, so $r' < 0$. On the other hand, if $q' < q$ then $r' = a - q'n = qn + r - q'n = r + (q - q')n \geq n$. We see that whenever $q' \neq q$ then r' does not satisfy the condition $0 \leq r' < n$. Therefore we must have $q' = q$ and therefore $r' = r$. q. e. d.

Proof analysis: The theorem has two parts: Existence ("there are") and uniqueness ("unique") of q, r. For the **proof of existence** we gave a way to **construct** q, r. To prove the property $r < n$ we used an **indirect proof.**

Here it was natural to construct the number q using an **extremal property** (the largest integer satisfying $qn \leq a$). We will develop this idea systematically in Chapter 10.

The **proof of uniqueness** is the direct translation of the initial ideas into mathematical language. This was also an indirect proof: We excluded the possibility of another representation than the one we first constructed.

Note that a may be negative, but n must be positive.[2] This will be useful later on. Pay special attention if a is negative:

Example

What is the remainder when dividing -10 by 3? The answer is in this[3] footnote.

The practical use of remainders

? Problem 8.1

a) *Suppose it is 11 o'clock now. What time will it be in 40 hours?*

b) *Suppose today is January 18, a Wednesday. Which day of the week will February 18th be?*

! Solution

a) After every 24 hours (one day) it will be the same time of day. So we need to find the remainder of $11 + 40 = 51$ modulo 24. That's three. The answer is 3 o'clock.
b) January has 31 days, so we have to advance by 31 days to get from January 18 to February 18. After every 7 days (one week) it will be the same day of the week. The remainder of 31 modulo 7 is 3, so February 18 is a (Wednesday+3) = Saturday. !

8.2 Congruences

Dividing the integers into evens and odd numbers is useful for many mathematical arguments. You saw examples of this in Problem 6.3 and in the proof that $\sqrt{2}$ is irrational in Section 7.2.

Even numbers leave the remainder 0 modulo 2, odd numbers leave the remainder 1. Instead of using remainders modulo 2 we could

[2] Initially you probably thought about $a > 0$. Convince yourself that the proof is also correct for $a \leq 0$.

[3] The remainder is 2, for $-10 = (-4) \cdot 3 + 2$. You could also write $-10 = (-3) \cdot 3 - 1$, but the remainder must always be ≥ 0.

classify integers according to their remainders modulo 3, or 4, or in general modulo n. This turns out to be so useful that there is a special notation for it. You will be astonished how a new notation, and a few simple facts about it, make certain difficult problems easy.

The following simple fact will be useful.

Lemma Let $a, b \in \mathbb{Z}$ and $n \in \mathbb{N}$. The following are equivalent:

(i) a and b leave the same remainder when divided by n.

(ii) $n|a - b$, that is, a and b differ by a multiple of n.

For example, 8 and 22 leave the same remainder when divided by 7 (namely 1) because their difference $22 - 8 = 14$ is divisible by 7. Conversely, the difference is divisible by 7 since the remainders are equal.

Try to prove the lemma yourself.

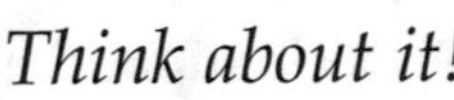

Proof.

Let $a, b \in \mathbb{Z}$ and $n \in \mathbb{N}$. Write (division by n with remainder)

$$a = qn + r, \quad b = pn + s \tag{8.1}$$

with quotients $p, q \in \mathbb{Z}$ and remainders $r, s \in \{0, 1, \ldots, n-1\}$.

Proof of (i) $\Rightarrow$ (ii): If $r = s$ then $a - b = (qn + r) - (pn + r) = (q - p)n$, so $n|a - b$ since $q - p \in \mathbb{Z}$.

Proof of (ii) $\Rightarrow$ (i): $n|a - b$ means that there is $k \in \mathbb{Z}$ satisfying $a - b = kn$. Then $kn = a - b = (q - p)n + (r - s)$ implies that $r - s$ is divisible by n. Because two different numbers $r, s \in \{0, \ldots, n-1\}$ have distance less than n it follows that $r = s$. q.e.d.

Proof analysis: This was a typical example of a **proof of equivalence.** In order to prove the equivalence of the statements (i) and (ii) about the numbers a, b, n, you fix a, b, n and prove that $(i) \Rightarrow (ii)$ and $(ii) \Rightarrow (i)$. Both of these proofs were direct proofs.

Equality of remainders is so important that there is a notation for it.

Definition Let $a, b \in \mathbb{Z}$ and $n \in \mathbb{N}$. We write

$$a \equiv b \mod n$$

(read: a **is congruent to** b **modulo** n) if $n | a - b$, or equivalently if a and b leave the same remainder when divided by n.[4]

Example

$22 \equiv 8 \mod 7$. Practically this means: In 22 days we have the same day of the week as in 8 days.

That's which day of the week, supposing today is a Wednesday? Because $8 \equiv 1 \mod 7$ and tomorrow (in 1 day) is a Thursday, it must be Thursday in 8 and in 22 days.

The numbers a, b may well be negative. For example, $-6 \equiv 1 \mod 7$ because $-6 - 1 = -7$ is divisible by 7: six days ago it was also a Thursday.

The definition implies that we always have $a \equiv r \mod n$ if r is the remainder of a modulo n. Conversely the remainder of a modulo n is the only number $r \in \{0, \ldots, n-1\}$ which satisfies $a \equiv r \mod n$.

You can calculate with congruences:

[4]This notion of congruence is not related to the geometric notion of congruence of two figures in the plane or in space, which means that one figure can be turned into the other by means of a translation, rotation or reflection.

Theorem **(Calculation rules for congruences)** Let $n \in \mathbb{N}$. Congruences mod n can be added, subtracted, multiplied and exponentiated. That is: If

$$a \equiv b \mod n$$
$$c \equiv d \mod n$$

then

$$\begin{aligned} a+c &\equiv b+d &&\mod n \\ a-c &\equiv b-d &&\mod n \\ ac &\equiv bd &&\mod n \\ a^k &\equiv b^k &&\mod n \end{aligned}$$

for all $k \in \mathbb{N}$.

Be careful! You *cannot divide* congruences. For example, we have $2 \equiv 6 \mod 4$ and $2 \equiv 2 \mod 4$, but $\frac{2}{2} \not\equiv \frac{6}{2} \mod 4$. See also Exercise E 8.11.

Proof.
By definition $a \equiv b, c \equiv d \mod n$ mean that $n|a-b$ and $n|c-d$. Then

$$n|a-b,\ n|c-d \Rightarrow n|(a-b)+(c-d) = (a+c)-(b+d)$$

which gives the first claim, and the second claim follows similarly. Now try to find a proof of the third claim (multiplication) yourself!

Think about it!

For the third claim write

$$ac - bd = ac - bc + bc - bd = (a-b)c + b(c-d)\,.$$

Now $n|a-b \Rightarrow n|(a-b)c$ and $n|c-d \Rightarrow n|b(c-d)$, and adding implies $n|(a-b)c + b(c-d) = ac - bd$, so $ac \equiv bd \mod n$.

If you use the third claim with $a = c, b = d$ then you get $a^2 \equiv b^2 \mod n$. Applying the same idea repeatedly you get $a^k \equiv b^k \mod n$ for all $k \in \mathbb{N}$. q. e. d.

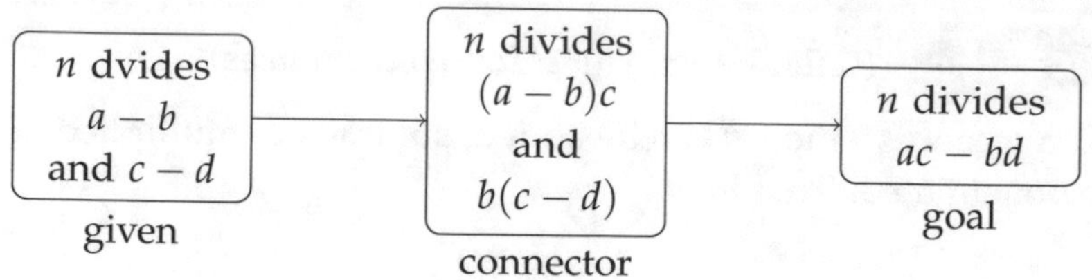

Figure 8.1 Solution scheme for the proof analysis

Proof analysis: This was a **direct proof**, and the main point was skilful calculations.

The transformations used in the proof for multiplication seem difficult to find. How on earth can you get this idea? This is a typical problem-solving situation. **What is our goal?** That $ac - bd$ is divisible by n. **What is given?** That $a - b$ and $c - d$ are divisible by n. In short: Goal: $ac - bd$. Data: $a - b$, $c - d$. **How can we connect the goal with the data?** The expression $ac - bd$ looks complicated, so let's begin with one part of it, the term ac. How can we connect this with the data? One way to do this is through the expression $(a - b)c = ac - bc$ since it has both $a - b$ and ac in it. What do we need to get from here to our goal? To get from $ac - bc$ to $ac - bd$ we need to add $bc - bd$, which equals $b(c - d)$. Summarizing, we have $ac - bd = (a - b)c + b(c - d)$. Figure 8.1 shows the solution scheme; it is an example of the general scheme in Figure 6.4.[5]

[5]This type of calculation, which is also called "clever addition of zero" (because we added $0 = -bc + bc$ between ac and bd) can be found in various places in mathematics. For example, in analysis, we add a clever form of zero when proving that the product of two convergent sequences converges to the product of their limits.

Examples

- What is the remainder of 100 modulo 7? You could calculate it directly, but this way is easier: From $10 \equiv 3 \mod 7$ we get $10^2 \equiv 3^2 \mod 7$, so

$$100 = 10^2 \equiv 3^2 = 9 \equiv 2 \mod 7.$$

So the remainder of 100 modulo 7 is 2. Cross-check: $100 = 14 \cdot 7 + 2$.

- What is the remainder of 2^{100} modulo 3? This would be very tedious to calculate directly. Here is a simple method: From $2^2 = 4 \equiv 1 \mod 3$ we get

$$2^{100} = 2^{2 \cdot 50} = (2^2)^{50} \equiv 1^{50} = 1 \mod 3\,.$$

So the remainder is 1.

- What is the last digit of the number 7^{77}? The last digit of a natural number (in the decimal system) is its remainder modulo 10. Let us try smaller exponents first: $7^2 = 49 \equiv 9 \equiv -1 \mod 10$. This is easy to exponentiate, for example $7^{76} = 7^{2 \cdot 38} = (7^2)^{38} \equiv (-1)^{38} = 1 \mod 7$, hence

$$7^{77} = 7^{76} \cdot 7 \equiv 1 \cdot 7 = 7 \mod 10\,.$$

So the last digit of 7^{77} is 7.

Exercises

E 8.1 If January 18, 2012 is a Wednesday, which day of the week is January 18, 2013? Formulate your solution using congruences. (Note that 2012 is a leap year.) 1

E 8.2 Explain how the well-known rule "even + even = even" can be reduced to the equation $0 + 0 = 0$ using congruences. Which equations correspond to the rules "even + odd = odd", "odd·odd = odd" etc.? 1

1-2 E 8.3 Prove $7|3^{105}+4^{105}$.

2 E 8.4 We write $\lg 2$ for the logarithm of 2 to the base 10, that is, the number which satisfies $10^{\lg 2}=2$. Prove that $\lg 2$ is irrational.

1-2 E 8.5 Prove that $6 \nmid n^3+5n+1$ for all $n \in \mathbb{Z}$.

3 E 8.6 Prove the following divisibility rules using congruences. All numbers are represented in the decimal system.

a) A natural number is divisible by 3 (resp. 9) if and only if the sum of its digits is divisible by 3 (resp. 9).

b) A natural number n is divisible by 7 if and only if the number n' obtained as follows is divisible by 7: Let a_0 be the last digit of n, so $n=10a+a_0$ where $a_0 \in \{0,\dots,9\}$. Then let $n'=a-2a_0$.

3 E 8.7 Find and prove a divisibility rule for division by 11.

3-4 E 8.8 Find more divisibility rules in the decimal system, for example for division by 13 or 17. Find a divisibility rule for division by 7 in the octal number system (number system with base 8).

2 E 8.9 Let $n \in \mathbb{N}$. Imagine writing down the remainders modulo n of $1^2, 2^2, \dots, (n-1)^2$. Are they all different? Why not?

2-3 E 8.10 (Quadratic residues)

a) Produce for each number $n=3,4,5$ a table in which each number $a=0,1,\dots,n-1$ is followed by the remainder of a^2 modulo n. (These remainders are called **quadratic residues** modulo n.)

b) Conclude that every square number has remainder 0 or 1 when divided by 4. What are the possible remainders modulo 3 and modulo 5?

c) Which remainders can the sum of two square numbers have modulo 4?

d) Are there square numbers of the form $3n-1$, $5n+4$, $7n+3$ (with $n \in \mathbb{N}$)?

e) Can 999,999 be written as the sum of two squares?

f) Prove: if $5|a^2+b^2+c^2$ with $a,b,c \in \mathbb{N}$ then $5|a$ or $5|b$ or $5|c$.

E 8.11 Two numbers $n, c \in \mathbb{N}$ are called **coprime** (or relatively prime) if 1 is the only natural number which divides both n and c. 2

a) How can you read off from the prime factorisations of n and c whether they are coprime?

b) Prove: if $n, a, c \in \mathbb{N}$ and n, c are coprime then $n|ac$ implies $n|a$.

c) Prove for $n, a, b, c \in \mathbb{N}$ satisfying $c|a$, $c|b$:

$$a \equiv b \mod n \Rightarrow \frac{a}{c} \equiv \frac{b}{c} \mod n, \quad \text{if } n, c \text{ are coprime.}$$

E 8.12 The aim of this exercise is to prove EUCLID's lemma. Be careful: in the proof we cannot use the uniqueness of the prime factorisation, since we used EUCLID's lemma to prove it. 3-4

a) (The case $p = 2$) Prove: if ab is even and b is odd then a is even. Argue carefully.

b) (The case $a < p$) Prove: if p is a prime number and if $a, b \in \mathbb{N}$ are such that $p \nmid b$ (p does not divide b) and $a < p$ then $p \nmid ab$.

c) Prove EUCLID's lemma using b).

E 8.13 Consider the following proposition and the given (wrong) "proof". 2

> **Lemma.** Every power 6^k with $k \in \mathbb{N}$ ends in the digit 6 when written in the decimal system.
>
> *Proof.* "Proof"
> In order to prove the claim, we show
>
> $$\forall k \in \mathbb{N} : 6^k \equiv 6 \pmod{10}.$$
>
> We will show this by contradiction.
>
> *Assumption:* $6^k \not\equiv 6 \pmod{10}$ for all $k \in \mathbb{N}$.
>
> If $6^k \not\equiv 6 \pmod{10}$ for each $k \in \mathbb{N}$ then this is true for $k = 2$ in particular. However,
>
> $$6^2 = 36 \equiv 6 \pmod{10}.$$
>
> This is a contradiction to the assumption, so the assumption must have been false. This proves the lemma.

Why is the proof wrong? Give a correct proof.

1-2 E 8.14 (Inverse of EUCLID's lemma) Let $p \in \mathbb{N}$, $p \geq 2$ be such that for all $a, b \in \mathbb{N}$ the following is true: If $p|ab$ then $p|a$ or $p|b$.

Show that then p is a prime number.

2 E 8.15 Prove: the number of positive divisors of a natural number n is odd if and only if n is a square number.

3 E 8.16 Let $m, n \in \mathbb{N}$ be coprime. Calculate

$$\lceil \frac{m}{n} \rceil + \lceil 2\frac{m}{n} \rceil + \cdots + \lceil (n-1)\frac{m}{n} \rceil$$

where $\lceil x \rceil$ is x 'rounded up', for example $\lceil 2.4 \rceil = 3$, $\lceil 5 \rceil = 5$.

2 E 8.17 Let a be a natural number. Decide for each of the propositions a) to d) below whether it implies the proposition

$$a \equiv 1 \pmod 4 \tag{8.2}$$

or whether it is implied by this proposition.

a) a is odd.

b) a is odd and can be represented as the sum of two positive square numbers.

c) $a^2 \equiv 1 \pmod 8$

d) $2a \equiv 2 \pmod 4$

Give a proof or a counterexample for each answer.

9 The pigeonhole principle

If many pigeons roost in few pigeonholes then there must be a hole with many pigeons.[1] This is so obvious that you may be surprised how many mathematical arguments are based on such reasoning, and what surprising consequences it has. It is an important tool for proofs of existence. The art is in recognising when and how it can be used. This is sometimes quite obvious, but sometimes hard to discern.

In this chapter we start with a precise statement of the pigeonhole principle and give a few examples from everyday life. Turning to mathematics, you will see how you can deduce some interesting facts about integers by considering remainders as pigeonholes. Then we will investigate the question how well any real number can be approximated by fractions, and see how the pigeonhole principle can be useful in a quite non-obvious way. Finally we apply the principle to a problem from graph theory and find that even in the greatest chaos there must be some order.

9.1 The pigeonhole principle, first examples

We start with the simplest version of the pigeonhole principle:

Pigeonhole principle

Let $n \in \mathbb{N}$. If $n+1$ pigeons roost in n holes then there must be a hole containing more than one pigeon.

This statement is optimal in the following sense: if there are only n pigeons in n holes then it is possible that there is only one pigeon in each hole.

[1]Nowadays when speaking of pigeonholes you usually think of open mail-boxes. So instead of pigeons you may think of envelopes if you wish. But it's more fun to think of birds, so we will stick to the original meaning.

D. Grieser, *Exploring Mathematics*, Springer Undergraduate Mathematics Series, https://doi.org/10.1007/978-3-319-90321-7_9

Proof.
If each hole contained at most one pigeon then there would be at most n pigeons in total. q. e. d.

Proof analysis: This is an **indirect proof.**

If you replace pigeons and holes suitably then you get interesting facts.

Examples

- Among any 13 people there are two whose birthdays are in the same month.

  (pigeons: people; holes: months)

  Important: 'there are two' always means: 'there are at least two'; there could also be three or more people having their birthdays in the same month.

- Among any three people there are two of the same sex.

  (pigeons: people; holes: sexes)

- Among the inhabitants of Liverpool there are two with the same number of hairs on their head.[2]

  (pigeons: inhabitants of Liverpool; holes: possible numbers of hairs, i.e. the numbers $0, 1, 2, \ldots, 300,000$. Put an inhabitant in hole i if he/she has i hairs on his/her head.)

The pigeonhole principle is very effective for **proofs of existence:** in the last example you know readily that *there are* two inhabitants with the same number of hairs. If you wanted to know *who* they are, you would need to count the hairs of many people – an almost impossible task. In this sense the pigeonhole principle is **non-constructive.**

Often we need a more general version of the pigeonhole principle.

[2]For this you need to know: 1. Every human has at most 300,000 hairs on his head (usually it's about 150,000). 2. the city of Liverpool has about 466,000 inhabitants (data from 2013).

General pigeonhole principle

Let $a, n \in \mathbb{N}$. If $an+1$ pigeons roost in n holes then there must be a hole containing more than a pigeons.

Again this is optimal: it would not be true with an or fewer pigeons.

Proof.
If each hole contained at most a pigeons then there would be at most an pigeons in total. q. e. d.

Examples

- There are 13 gifts under the Christmas tree, for 3 children. Then at least one child gets at least 5 gifts.

  (pigeons: gifts; holes: children. The argument is: if every child got at most 4 gifts then there would be at most $3 \cdot 4 = 12$ gifts in total. Here $n = 3$ and $a = 4$.)

- Among the inhabitants of England there are at least 170 with the same number of hairs on their heads.[3]

  (There are more than $51,000,170 = 170 \cdot 300,001$ inhabitants, so it cannot be that each number of hairs occurs only at most 170 times.)

In these examples it was easy to see how to use the pigeonhole principle. In the following problems you need to think carefully about what you could take as pigeons and what as holes.

A useful rule of thumb is: the objects whose (multiple) existence you want to prove are the pigeons, their properties are the holes.

? Problem 9.1

Some people are in a room. Show that there are two among them who are acquainted with the same number of people in the room.[4]

---

[3] By the 2011 census England has about 53,000,000 inhabitants.

[4] We assume that the 'acquaintance' relation is symmetric, that is: if A is acquainted with B then B is acquainted with A.

Investigation

▷ Let $n \geq 2$ be the number of people in the room. The statement 'there are two ...' suggests that we use the pigeonhole principle. **What could be the pigeons, what the holes?** We want to prove the existence of two *people*, the relevant property is the *number of acquaintances*. So the pigeons should be the people, the holes the possible numbers.

▷ Which holes are there? A person in the room can have $0, 1, \ldots$, or $n-1$ acquaintances in the room. These numbers are the holes.

▷ So we have n pigeons and n holes. The pigeonhole principle is not applicable. What now?

▷ Look again: if one person knows everyone else (thus has $n-1$ acquaintances) then there can be no person having 0 acquaintances. So either hole 0 or hole $n-1$ is empty.

▷ Therefore only $n-1$ holes can be used, so one of them must contain 2 pigeons.

You could write this up as follows:

Solution of Problem 9.1

Let $n \geq 2$ be the number of people in the room. We number them $1, \ldots, n$. Let b_i be the number of acquaintances of person i. Then $b_i \in \{0, 1, \ldots, n-1\}$ for each i. If there is a person i with $b_i = n-1$ (a 'socialite') then she knows everyone else, so there can be no person j having $b_j = 0$. Therefore at most $n-1$ different numbers appear among $b_1, \ldots, b_n$, so there must be $i \neq j$ with $b_i = b_j$.

Remark

If you represent the people as vertices of a graph and the acquaintanceships as edges then you can state this result in the language of graphs:

In any graph[5] *there are two vertices having the same degree.*

9.2 Remainders as pigeonholes

Sometimes it is useful to use remainders modulo n as pigeonholes. Here is a simple example.

? Problem 9.2

Given 10 natural numbers, show that there are two among them whose difference is divisible by 9.

The goal is divisibility by 9, this suggests that we use remainders modulo 9 as holes.

! Solution

There are 9 possible remainders when dividing by 9: $0, 1, \ldots, 8$. Therefore two of the 10 given numbers have the same remainder modulo 9. Their difference is divisible by 9. !

The analogous statement with 10 replacing 9 would be: among 11 natural numbers there are two that end in the same digit. This is obvious: there are only 10 possible digits. Note that the last digit is the remainder modulo 10.

The pigeonhole principle can also be useful for proving the existence of a single object with certain properties:

? Problem 9.3

Prove that among the powers $7, 7^2, 7^3, \ldots$ *there is one ending in 001.*

Investigation

What are we looking for? Three final digits. That's the same as the remainder modulo 1000. So we could try to use the pigeonhole principle where the holes are the remainders modulo 1000. **What should be the pigeons?** The objects whose existence we want to prove: the powers of 7. So we get two powers of 7 having the same remainder modulo 1000. How can we use this for the problem?

---

[5]with at least two vertices, and without double edges or loops

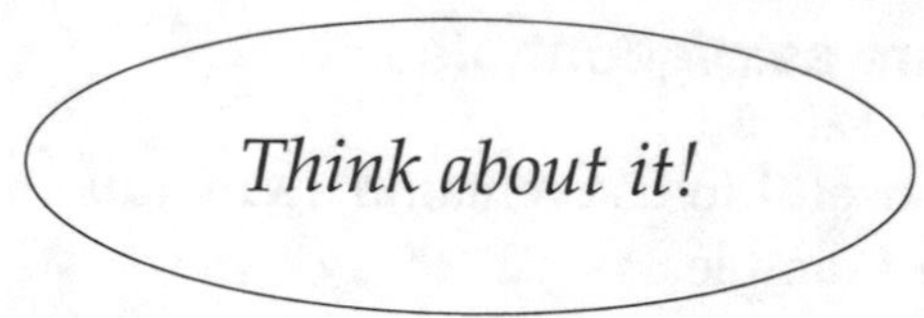

We proceed **step by step**. Let us write it down: $7^k \equiv 7^l \mod 1000$, or equivalently $1000|7^k - 7^l$ for two different natural numbers k, l. How can we use this? We could take out the smaller power: let's say $k > l$, then $7^k = 7^l \cdot 7^{k-l}$, so $7^k - 7^l = 7^l \cdot (7^{k-l} - 1)$. If this is divisible by 1000 then $7^{k-l} - 1$ must also be divisible by 1000, so 7^{k-l} ends in 001, done!

This was almost perfect, but let's write it down with all details.

Solution of Problem 9.3

A number ends in 001 if and only if it has the remainder 1 modulo 1000. Therefore we consider the remainders modulo 1000 as holes. There are 1000 remainders: $0, 1, \ldots, 999$. Therefore, among the 1001 numbers

$$7^1, 7^2, 7^3, \ldots, 7^{1001}$$

there must be two, say 7^k and 7^l where $k > l$, having the same remainder. Then

$$1000|7^k - 7^l = 7^l(7^{k-l} - 1).$$

Now $1000|7^l(7^{k-l} - 1)$ implies $1000|7^{k-l} - 1$ since the prime factorisation of $7^l(7^{k-l} - 1)$ must contain $1000 = 2^3 \cdot 5^3$, and since 7 is prime, this can only be contained in the prime factorisation of $7^{k-l} - 1$. Therefore, 7^{k-l} leaves the remainder 1 modulo 1000, so it ends in 001.

Review of Problem 9.3

The problem suggested using remainders modulo 1000 as holes and powers of 7 as pigeons. It was less clear what use it would be to have two powers with the same remainder. What saved us was the fact that $7^k = 7^l \cdot 7^{k-l}$.

The proof shows that among the numbers $7, 7^2, \ldots, 7^{1000}$ there must already be one ending in 001.

An analogous statement holds for any number other than 7 which is not divisible by either 2 or 5. And instead of 001 we can require n zeroes followed by 1, for any n. The proof is the same.

9.3 An exploration: approximation by fractions

We want to explore the following question.

> **Approximation problem:** *How well can an arbitrary real number be approximated by fractions?*

Your first reaction might be: what a strange question, of course any real number a can be approximated *arbitrarily well* by fractions:[6] Just cut off the decimal representation of a after a few digits. For example, $\pi = 3.1415\ldots$ can be approximated by $3.141 = 3141/1000$ with an error of $0.0005\ldots$. If you want a better approximation then simply use more digits.[7]

We could stop here: problem solved, exploration finished. Or we could try to discover more, taking our vague question only as a starting point.

Let us stick to the example $\pi = 3.1415926\ldots$. If the fraction m/n is to be close to π then $n\pi$ should be close to an integer. So let us look at some multiples of π and their distance to the nearest integer, see Table 9.1:

Surprise: 7π is much closer to an integer (22) than all the other listed multiples of π. So $22/7$ will be a very good approximation of π, considering that the denominator is quite small. We calculate $22/7 = 3.142\ldots$ and see that the error is about 0.001. Compare this with our first idea of cutting off the decimal expansion. Table 9.2 shows some of these fractions and their approximation errors. It also shows another very good approximation, $355/133$. What is remarkable about this table? The fractions $22/7$ and $314/100$ approximate π about equally well, but $22/7$ has a much smaller numerator and denominator.

[6]By **fraction** we always mean a fraction of integers; numbers that can be written as fraction of integers are called **rational.** In what follows we do not distinguish between a fraction and the rational number represented by it.

[7]Maybe you are also wondering: Why is this question in the chapter on the pigeonhole principle? Just wait. Let the question, not the method, be our guide.

n	$n\pi$ (rounded)	distance
1	3.14	0.14
2	6.28	0.28
3	9.42	0.42
4	12.57	0.43
5	15.71	0.29
6	18.85	0.15
7	21.99	0.01
8	23.13	0.13
9	28.27	0.27
10	31.42	0.42

Table 9.1 Distance of $n\pi$ to the nearest integer, rounded to 2 digits

fraction			error
$^{31}/_{10}$	$=$	3.1	0.04...
$^{314}/_{100}$	$=$	3.14	0.001...
$^{3141}/_{1000}$	$=$	3.141	0.0005...
$^{22}/_{7}$	$=$	3.1428...	0.001...
$^{355}/_{113}$	$=$	3.1415929...	0.0000003...

Table 9.2 Approximations of π by fractions

This is even more extreme with $^{355}/_{113}$. Although the denominator of this fraction is only a little bigger than that of $^{314}/_{100}$, it gives a much better approximation.

We see that interesting things happen if we take the size of the denominator seriously in our investigation.[8] So we make our problem more specific:

> **Approximation problem, version 2:** *Investigate how the* ***approximation error*** *is related to the* ***size of the denominator*** *when we approximate a real number a by fractions.*

So far we have only considered the example $a = \pi$. A special property of this number is that it is **irrational**, that is, it cannot be written

[8] Instead we could take the size of the numerator. This would not make much of a difference since $a \approx {}^{m}/_{n}$ implies $m \approx an$.

precisely as a fraction.[9] But even for rational numbers a the problem is interesting. For example, consider $a = 0.51$: Of course this is precisely $51/100$, but you can approximate it by the much simpler fraction $1/2$, up to the quite small error 0.01.

Our question also has practical relevance: suppose you want to construct a transmission – for example, on a bicycle – with two gears which has a transmission ratio close to a given value a. You want your gears to have few teeth, in order to save money. If the gears have m and n teeth then the transmission ratio will be m/n. So you will need to balance the cost (which depends on $m + n \approx (a+1)n$, so on n) against the error $|a - m/n|$, and this is precisely our problem above.

Version 2 of the problem is still a little vague. Let us state a few concrete questions. First we fix some notation:

- a: a real number, assumed positive for simplicity
- $n \in \mathbb{N}$: the denominator in an approximating fraction
- $\varepsilon > 0$: the approximation error[10]

We will denote the following proposition by $\text{Approx}(a, n, \varepsilon)$.

'a can be approximated by a fraction of denominator n with an error of at most ε'

That is: 'There is an $m \in \mathbb{N}_0$ satisfying $\left|a - \frac{m}{n}\right| \leq \varepsilon$'

Here are a few questions that we could ask.

Approximation problem, version 3:

1. *Given a and n, what is the minimal ε for which* $\text{Approx}(a, n, \varepsilon)$ *is true?*
2. *Given a, are there certain values of n for which* $\text{Approx}(a, n, \varepsilon)$ *is true with very small ε?*
3. *Given n, what is the minimal ε for which* $\text{Approx}(a, n, \varepsilon)$ *holds for all a?*

---

[9] This is not so easy to prove. See (Hardy and Wright, 2008) for example.

[10] The Greek letters ε, δ (epsilon and delta) are frequently used in mathematics for small positive numbers.

Maybe you can think of more questions?

Note the difference between the third question and the other two: In 1. and 2. the number a is fixed at the beginning, and ε may depend on a. In 3. the integer n is fixed, and we are asking for an ε which works for all a. In both cases ε will depend on n.

To understand what happens we make a sketch, see Figure 9.1. We draw the points $0, \frac{1}{n}, \frac{2}{n}, \ldots$ on the real line. Then a will lie between two such points, or coincide with one of them.

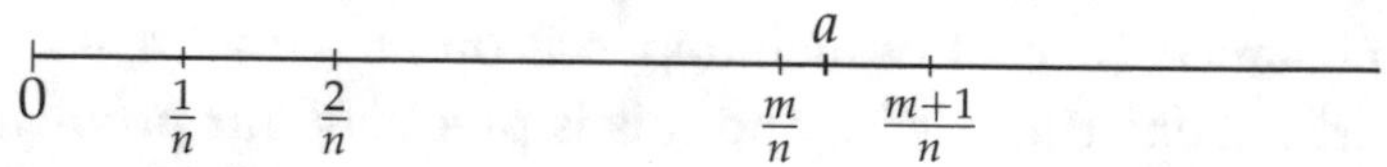

Figure 9.1 Approximation of a by the fraction $\frac{m}{n}$

We can read off that $\varepsilon = \frac{1}{2} \cdot \frac{1}{n} = \frac{1}{2n}$ always works, but also that smaller values, even $\varepsilon = 0$, might work, depending on a. This is all we can really say on Question 1.

Now let us check Question 3: which ε will work for *all* a? As we saw, $\varepsilon = \frac{1}{2n}$ works for all a; but if a lies half way between two of the marked points, for example if $a = \frac{1}{2n}$, then no smaller value of ε will work for it. So the answer to Question 3 is: the smallest such ε is $\varepsilon = \frac{1}{2n}$.

Question 2 is much more interesting, as we saw in the case $a = \pi$. It is still not a very precise question. What does "very small ε" mean? Of course very large values of n will allow very good approximations, so ε will depend on n. We saw that $\varepsilon = \frac{1}{2n}$ works for all n. But in the case $a = \pi$ the approximation errors of $22/7$ and of $355/113$ are much smaller than $1/14$ and $1/226$, respectively. Can we find such especially good approximations for any a? How good can they possibly be?

We found the good approximation $22/7$ of π by considering the distance of $n\pi$ to the nearest integer. So let us investigate this in general: how small can we make this distance when we allow numbers n up to a certain size?

Problem 9.4

Let $N \in \mathbb{N}$. Find the smallest number δ, depending on N, having the following property: for any real number $a > 0$ at least one of the numbers $a, 2a, \ldots, Na$ lies within distance δ of an integer.

This problem seems to be closer to Question 3 than to Question 2 since the same δ is to work for all a. However, we will see that the solution will help us in investigating Question 2. Clarify for yourself the difference between this problem and Question 3.

Investigation

▷ We need to understand the problem well. The statement is quite complicated: ... smallest ... any ... at least one .... Look at each of these words to understand what we are looking for. Consider some examples of a and N and find the best possible δ.

▷ We look at a **special case**: what is the answer for $N = 1$? For which δ is it true that any a lies within δ of an integer?
Clearly this is true for $\delta = 1/2$ but not for smaller values of δ (take $a = 1/2$). So the answer for $N = 1$ is $\delta = 1/2$.

▷ How about $N = 2$? Now we need to consider both a and $2a$. Play with this, try out different a, try to construct the worst case.

Think about it!

▷ It is time to **introduce some new notation**. For $x \in \mathbb{R}$, $x \geq 0$ denote by $\text{frac}(x)$ the fractional part of x.[11] Here is an exact definition.

Definition For $x \in \mathbb{R}$, $x \geq 0$ let $\lfloor x \rfloor$ (**Gauss bracket** of x) be the largest integer which is less than or equal to x, and let $\text{frac}(x) = x - \lfloor x \rfloor$.

Examples: $\lfloor 3.1 \rfloor = 3$ and $\text{frac}(3.1) = 0.1$. $\lfloor 5.9 \rfloor = 5$ and $\text{frac}(5.9) = 0.9$. $\lfloor 3 \rfloor = 3$ and $\text{frac}(3) = 0$.
We always have $0 \leq \text{frac}(x) < 1$.

[11] Instead of $\text{frac}(x)$ the notation $\{x\}$ is often used. We use $\text{frac}(x)$ in order to avoid confusion with the set having x as its only element.

▷ The fractional part of a real number determines its distance to the nearest integer. Therefore we consider $\text{frac}(a), \text{frac}(2a)$. Maybe you noticed that δ cannot be smaller than $1/3$ (for $a = 1/3$ or $a = 2/3$). So we **conjecture** that for $N = 2$ the optimal δ is $\delta = 1/3$. Let us try to show this.

▷ We want to show: for any a at least one of the numbers $\text{frac}(a)$, $\text{frac}(2a)$ lies in one of the outer intervals, shown thicker, in Figure 9.2. Here we include the endpoints $1/3$, $2/3$ in the intervals.

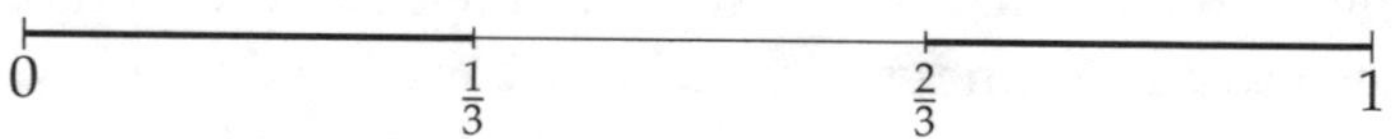

Figure 9.2 Three thirds for $\text{frac}(a), \text{frac}(2a)$

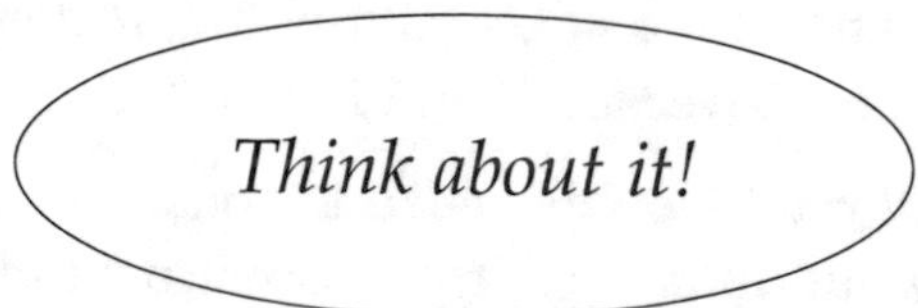

▷ What could we conclude if this was wrong? (We try a **proof by contradiction**.) Then both $\text{frac}(a)$, $\text{frac}(2a)$ would have to lie in the middle third. Then the distance between them would be at most $1/3$.[12]

▷ But if the distance between $\text{frac}(2a)$ and $\text{frac}(a)$ is at most $1/3$ then $2a - a = a$ must lie within $1/3$ of an integer, so $\text{frac}(a)$ would lie in one of the outer thirds. Contradiction!

▷ We need to check something here: is it true that the distance from $b - a$ to the nearest integer is at most $|\text{frac}(b) - \text{frac}(a)|$? (In our case $b = 2a$.) This looks reasonable, but let's postpone the details and first consider general N.

Try to generalize the argument to general N. What could be the optimal δ? How could you prove that? If you are not ready for the general case yet, try $N = 3$ first.

[12] In fact it is less than $1/3$, but this does not matter for the proof.

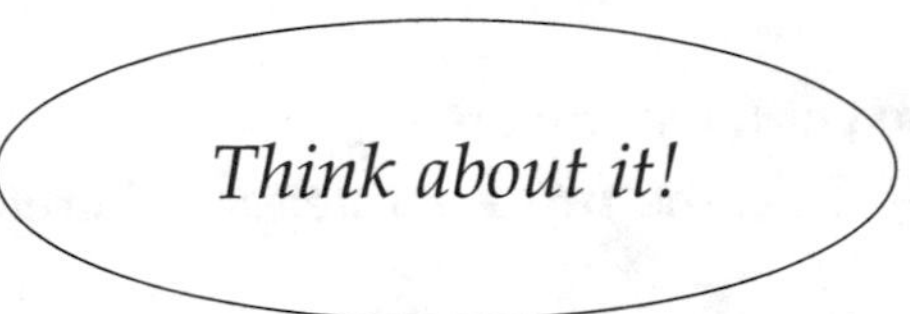

▷ For $N = 3$ we divide the interval $[0, 1]$ into four quarters. If none of the three numbers $\text{frac}(a)$, $\text{frac}(2a)$, $\text{frac}(3a)$ lies in the leftmost or rightmost quarter then they must all lie in the two remaining quarters, so by the pigeonhole principle two of them must lie in the same quarter. Then their distance would be less than $1/4$, and we get a contradiction as before. Therefore $\delta = 1/4$ works. A smaller δ would not work since for $a = 1/4$ each of the numbers $1/4$, $2/4$, $3/4$ has distance $\geq 1/4$ to the nearest integer.

▷ For $N = 1$ we got $\delta = 1/2$, for $N = 2$ we got $\delta = 1/3$ and for $N = 3$ we obtained $\delta = 1/4$. This cries out for a conjecture: the best possible δ is $1/(N+1)$.

! Solution of Problem 9.4

Claim: The smallest δ in Problem 9.4 is $\delta = \frac{1}{N+1}$.

For the proof we need a lemma.

Lemma Let $x, y \in \mathbb{R}$, $x, y \geq 0$. Then $|\text{frac}(x) - \text{frac}(y)| \leq \delta$ implies that $x - y$ lies within δ of an integer.

Proof.
Write $x = \lfloor x \rfloor + \text{frac}(x)$, $y = \lfloor y \rfloor + \text{frac}(y)$. Then

$$x - y = (\lfloor x \rfloor - \lfloor y \rfloor) + (\text{frac}(x) - \text{frac}(y)) .$$

The first parenthesis on the right is an integer, the second one has absolute value less than or equal to δ by assumption. The lemma follows. q. e. d.

We now prove the claim above, which solves Problem 9.4.

Proof.
For $a = \frac{1}{N+1}$ we get the numbers $\frac{1}{N+1}, \ldots, \frac{N}{N+1}$. Each of them differs by at least $\frac{1}{N+1}$ from the nearest integer. Therefore we must have $\delta \geq \frac{1}{N+1}$.

Now let $\delta = \frac{1}{N+1}$. Let $a > 0$ be arbitrary. We show that at least one of the numbers $a, 2a, \ldots, Na$ lies within δ of an integer. Consider the $N + 1$ intervals[13]

$$I_0 = [0, \delta], \ I_1 = [\delta, 2\delta], \ \ldots, \ I_N = [N\delta, 1].$$

Each interval has length δ, including the last one since $1 = (N+1)\delta$. Suppose none of the numbers na, $n = 1, 2, \ldots, N$ lies within δ of an integer. Then none of the numbers $\text{frac}(na)$ lies in I_0 or in I_N. Therefore, all of these N numbers lie in the remaining $N - 1$ intervals. By the pigeonhole principle at least two of them, say $\text{frac}(ka)$ and $\text{frac}(la)$ with $k > l$, must lie in the same interval. This implies $|\text{frac}(ka) - \text{frac}(la)| \leq \delta$. Then the lemma implies that $ka - la$ lies within δ of an integer. Now $k, l \in \{1, \ldots, N\}$, $k > l$ imply $k - l \in \{1, \ldots, N-1\}$, so $ka - la = (k-l)a$ is one of the numbers $a, 2a, \ldots, Na$. This contradicts our assumption. Therefore the assumption was wrong and the claim is proven. q. e. d.

!

Proof analysis: The proof that $\delta = \frac{1}{N+1}$ works was a **proof by contradiction:** the assumption that the claim is wrong for some a led to a contradiction. The pigeonhole principle could be applied since from two numbers in the same 'hole' a new number (their difference) could be constructed, which provided the contradiction.

In order to show that smaller values of δ don't work it sufficed to give an example a. (Recall that the negation of a 'for all' proposition is a 'there exists' proposition.)

Let us return to Question 2 in the third version of the approximation problem: given a, are there certain special values of n and corresponding m for which $\left|a - \frac{m}{n}\right| < \varepsilon$ holds with very small ε? We already

[13] It is irrelevant that some of the intervals intersect at their endpoints. What matters is that their union contains the interval $[0, 1)$, the set of possible values of $\text{frac}(a)$.

know that $\varepsilon = \frac{1}{2n}$ always works, for any n; but we are aiming for much smaller ε, generalizing the example of $a = \pi$ and $\frac{m}{n} = \frac{22}{7}$. We begin by restating the solution of Problem 9.4 as follows:[14] Let $a > 0$.

Given $N \in \mathbb{N}$ there are $n \in \mathbb{N}$, $n \leq N$, and $m \in \mathbb{N}_0$ with

$$|na - m| \leq \frac{1}{N+1}. \tag{9.1}$$

Dividing by n we get $\left|a - \frac{m}{n}\right| \leq \frac{1}{n(N+1)}$. For $N \geq 2$ the error $\frac{1}{n(N+1)}$ is already better than the error $\frac{1}{2n}$. Still this is not quite satisfactory since we would like to express (or estimate) the error in terms of the denominator n. That's easy: From $n \leq N$ we get $\frac{1}{N+1} \leq \frac{1}{n+1}$ and then

$$\left|a - \frac{m}{n}\right| \leq \frac{1}{n(n+1)}. \tag{9.2}$$

But this isn't quite what we wanted either, since all we know is that there is *some* n (and then a corresponding m) for which this inequality is true. If n was, say, 2 then that's not a very good approximation! However, there is a better way to use (9.1), which yields:

Theorem Let $a > 0$ be irrational. Then there are infinitely many fractions $\frac{m}{n}$ for which inequality (9.2) holds.

Proof.
We show that given any finite set of fractions there is another fraction which is not in the set and which satisfies inequality (9.2). Starting with the empty set we get arbitrarily many fractions satisfying the inequality. Therefore there cannot be only a finite number of them.

Thus, let M be a finite set of fractions. For each of these fractions $\frac{m}{n}$ consider the number $|na - m|$. It is positive, not zero, since a is irrational. So we get finitely many positive numbers. Let δ_0 be the smallest among them. (If M is the empty set then choose δ_0 to be any positive number.) Now we choose a natural number N satisfying $\frac{1}{N+1} < \delta_0$ and then, using (9.1), find $n' \leq N$ and m' satisfying $|n'a - m'| \leq \frac{1}{N+1}$. We use this inequality in two ways: First, it implies

[14] This is known as **Dirichlet's approximation theorem.**

$|n'a - m'| < \delta_0$, hence $\frac{m'}{n'} \notin M$ by definition of δ_0. Second, we divide by n', and using $n' \leq N$ we get $\left|a - \frac{m'}{n'}\right| \leq \frac{1}{n'(n'+1)}$, which was to be shown.

q. e. d.

> *Proof analysis:* This was a **direct proof.** When constructing another fraction the **extremal principle** was useful ('Let δ_0 be the smallest ...'). We will introduce this principle systematically in Chapter 10.

What does the theorem buy us? First, note that for each $n \geq 2$ there can be at most one m satisfying (9.2) (why?). Therefore, the theorem implies that the estimate is true for infinitely many n, hence for any given $\varepsilon > 0$ we can find such n with $\frac{1}{n(n+1)} < \varepsilon$, and then $\text{Approx}(a, n, \varepsilon)$ holds.

But there is much more to it: the estimate (9.2) is much better than the approximation obtained by cutting off the decimal expansion. If you cut off after the kth digit after the decimal point, then the approximating fraction has the form $\frac{m}{10^k}$, and the error is at least $\frac{1}{10^{k+1}}$ unless the $(k+1)$st digit happens to be zero. So here we have $n = 10^k$, and the error is of the order of $\frac{1}{10n}$. For $k > 1$ this is much bigger than the error $\frac{1}{n(n+1)}$ in (9.2): The former has denominator linear in n and the latter quadratic.

More remarks on the approximation problem

- The statement of our last theorem also holds for rational a but is uninteresting then: write $a = \frac{p}{q}$ and choose $m = kp$, $n = kq$ for $k = 1, 2, \ldots$. However, the question of good approximation by fractions with small denominators remains interesting for rational numbers, as we saw. The theory of **continued fractions** gives an effective algorithm for finding such fractions (in the case of rational or irrational a). See for example (Hardy and Wright, 2008) or (Silverman, 2012).

- The approximation (9.1) and the subsequent theorem hold for all $a > 0$. However, for certain numbers a there are much better approximations. For example, for some a the error $\frac{1}{n(n+1)}$ in (9.2) can be replaced by $\frac{1}{n^3}$, which means a much better approximation for large n. On the other hand it is known that such an improvement is not possible for the golden ratio, $a = \frac{1+\sqrt{5}}{2}$. So this number is only badly approximable.

This circle of questions is called **diophantine approximation.** It plays an important role in the theory of dynamical systems. For example, around 1900 Poincaré was led to such questions when investigating the motion of the planets.

❑ **The gap problem:** The fact stated in Problem 9.4 has other interesting consequences. Consider Figure 9.3. A goblin walks on a circular path of length 1 with constant step size a. There is a gap in the path of length $\varepsilon > 0$. Then the following is true: if a is irrational then the goblin will fall into the gap at some point – no matter how small ε is and where he starts.[15]

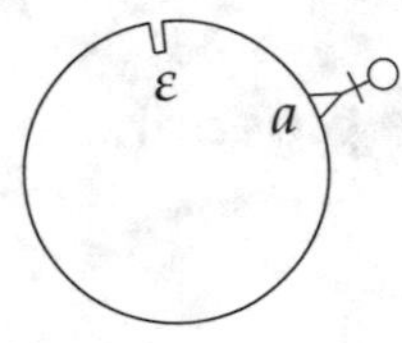

Figure 9.3 If a is irrational then the goblin will fall into the gap eventually

Proof.
Mark the starting point by 0 and measure lengths in the walking direction. Since the path has length 1 the steps are at frac(a), frac$(2a)$, frac$(3a)$,.... Choose $N \in \mathbb{N}$ satisfying $\frac{1}{N} < \varepsilon$. By Problem 9.4 there is an $n \leq N$ such that frac$(na) < \frac{1}{N}$ or frac$(na) > 1 - \frac{1}{N}$. We consider the first case, the argument in the second case is similar. Let $b =$ frac(na). Since a is irrational we have $b > 0$. The first n steps move the goblin from 0 to b. After another n steps it will have moved to $2b$, after the next n steps to $3b$ etc. Now $|b| < \frac{1}{N} < \varepsilon$, so one these numbers $b, 2b, \ldots$ falls into the gap. q. e. d.

In mathematical language we have proved: Let a be irrational. Then for each interval $[x, y] \subset [0, 1]$, where $x < y$, there is a $k \in \mathbb{N}$ so that frac$(ka) \in [x, y]$. This is usually expressed as follows:

[15] We assume that the goblin's feet are infinitely thin.

Theorem Assume a is irrational. Then the set of numbers $\mathrm{frac}(ka)$, $k \in \mathbb{N}$, is dense in $[0,1]$.

Convince yourself that this is not true if a is a rational number.

The sequence $\mathrm{frac}(ka)$, $k = 1, 2, 3, \ldots$, is a simple example of an *ergodic dynamical system.*

In Exercise E 9.15 you find a pretty application to powers of integers.

9.4 Order in chaos: the pigeonhole principle in graph theory

? Problem 9.5

Prove that among any 6 people there are 3 who know each other or 3 who don't know each other.

It is useful to represent this by a graph with coloured edges. The vertices of the graph correspond to the people. Any two vertices are connected by an edge, which is coloured red (continuous line) or green (dashed line). Red means that the two people know each other, green means they don't. See Figure 9.4 for an example.

We want to show that there is a monochromatic triangle (i.e. three vertices such that all three edges between them have the same colour), *no matter how the edges are coloured.*

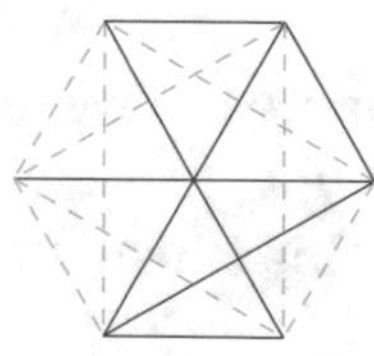

Figure 9.4 A graph with edges coloured by two colours

Play with this to see it is actually true. Try to prove it. Hint: Look at all the edges emanating from one vertex. What can you say about them?

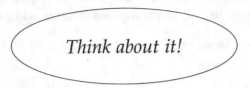

Think about it!

! Solution of Problem 9.5

We solve the problem translated into graph language as explained. Let V be one of the vertices. There are 5 edges emanating from it. By the pigeonhole principle three of them must have the same colour, say red. Say the edges from V to V_1, to V_2 and to V_3 are red. Now look at the edges between V_1, V_2 and V_3. If one of them is red then its two endpoints and V form a red triangle. Otherwise all of them are green, so they form a green triangle. !

Does this also work with 5 vertices (people)? No: draw all outer edges of a pentagon red and all diagonals green. Then there is no monochromatic triangle.

Let us generalize the question: is it true that among sufficiently many people there must be 4 who know each other, or 4 who don't know each other (i.e. none of the 4 knows any other)? In the language of graphs this would correspond to 4 vertices so that all 6 edges between them have the same colour. We call this a complete monochromatic quadrilateral.

It is a little simpler to first consider an intermediate case: colour all edges between 10 vertices red or green. Then there is a red triangle or a complete green quadrilateral. Prove this!

Is 10 the smallest number for which this is true? No. It turns out that 9 also works, but this is a little harder to prove. For 8 vertices this is not true. We symbolise this fact by the equation $R(3,4) = 9$ and say that 9 is the **3,4-Ramsey number.** Similarly, we showed that $R(3,3) = 6$. The higher Ramsey numbers are difficult to find precisely, for example, $R(5,5)$ is not known.

We can generalize the question further by using more than two colours. One can show: Let $k, a_1, \ldots, a_k \in \mathbb{N}$. Then there is a number R so that every colouring of all edges between R points with the colours $1, 2, \ldots, k$ has a complete 1-coloured a_1-gon or a complete 2-coloured a_2-gon or $\ldots$ or a complete k-coloured a_k-gon. That is:

If there is enough chaos then you will find order in it.

9.5 Toolbox

The pigeonhole principle is a tool which you should always have in mind when you want to **prove existence** of something. It is not always obvious how to use the pigeonhole principle. You need to **think about what could be the pigeons and the holes.** If you are looking for objects with certain properties then the objects will usually correspond to the pigeons and the properties to the holes. In problems about integers **remainders** are often good candidates for the holes.

The pigeonhole principle yields the existence of *several* objects with certain properties. Sometimes it is also useful if you only need *one* object, for example when you can use two objects to create another one using algebraic operations, e.g. difference or quotient (Problems 9.3 and 9.4).

Exercises

1 E 9.1 At least how many children must be in a room so you can be sure that three of them have their birthday in the same month?

1 E 9.2

a) There are five pairs of green socks and five pairs of red socks in a drawer. You cannot tell left socks from right socks. At least how many socks do you need to take blindly out of the drawer (without putting them back) if you want to be sure to have a matching pair?

b) In a shoe cabinet there are 10 different pairs of shoes. At least how many shoes do you need to take blindly out of the cabinet (without putting them back) if you want to be sure to have a matching pair? How many do you need to take if you have five identical brown pairs and five identical black pairs?

E 9.3 Suppose you walk around on a field of snow. Prove that at some point you have to step into your own footsteps (at least partially). How soon will this happen? Which quantities do you need, to find a number of steps by which you will certainly have done this? 1-2

E 9.4 Prove using the pigeonhole principle: Let a triangle and a line be given. Then the line intersects at most two sides of the triangle. 1-2

E 9.5 Show that you cannot cover an equilateral triangle completely by two smaller equilateral triangles. 2

E 9.6 In Figure 9.4 there are three monochromatic triangles, not just one. Is this always the case? Or else are there always at least two such triangles? 2-3

E 9.7 Let $a \in \mathbb{N}$ and let $x \in \mathbb{Z}$ be a number which is not divisible by 7. Consider the numbers $a, a+x, a+2x, \ldots, a+6x$. Does there have to be one among them which is divisible by 7? 2

E 9.8 Let $n \in \mathbb{N}$ and $a_1, \ldots, a_n \in \mathbb{N}$. Prove that there are indices j and k with $j \leq k$ so that $a_j + \ldots + a_k$ is divisible by n. 2-3

E 9.9 Prove that among any 52 *different* numbers from the set $\{0, \ldots, 99\}$ there must be two whose sum is 100. 2-3

E 9.10 Prove that among any 52 natural numbers there are two whose sum or difference is a multiple of 100. (Note that 0 is also a multiple of 100.) 3

E 9.11 A **lattice point** in the plane is a point whose coordinates are integers. Imagine a tree standing at every lattice point except the origin, see Figure 9.5. Standing at the origin, we look in some direction. Prove: 2-3

a) If all 'trees' have thickness zero then you will see a tree in all directions of rational slope, and you will see no tree in directions of irrational slope.[16]

[16]The slope of a ray starting at zero is defined as b/a if (a, b) is any point on the ray except the origin. For vertical rays the slope is defined to be ∞. For the purposes of this problem we count ∞ as rational.

b) If each tree has thickness $\varepsilon > 0$ then we will see a tree in every direction. That is, every ray starting at the origin hits a tree, no matter how small ε is.

c) Find a value d_ε, depending on ε, for which you can be sure that in b) you will see a tree in every direction at distance at most d_ε.

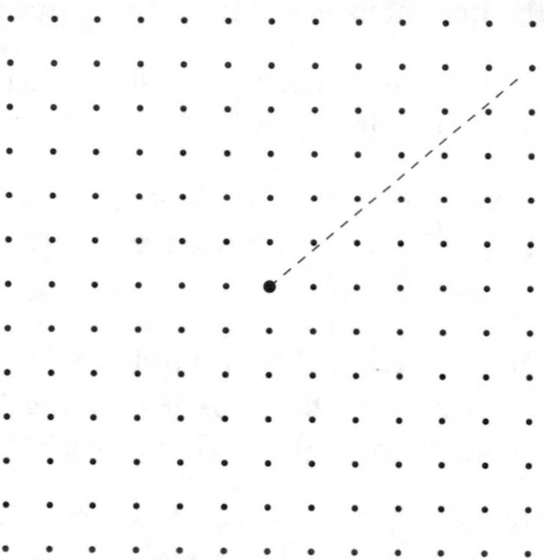

Figure 9.5 Trees obstruct the view

2 E 9.12 What is the largest number of lattice points which you can choose to satisfy the following condition: the midpoint between any two of the chosen points is not a lattice point.

3 E 9.13 Let $a, b \in \mathbb{N}$ be coprime, that is, the only natural number c dividing both a and b is $c = 1$. Prove that there are $n, m \in \mathbb{N}$ satisfying $an - bm = 1$.

3 E 9.14 Prove that there is a FIBONACCI number which is divisible by 1000.

3 E 9.15 Prove that there is a power of 2 whose decimal representation starts with the digits 999999. Also show that the same is true if 2 is replaced by any natural number n which is not a power of 10.

3 E 9.16 Let $n \in \mathbb{N}$. At least how many numbers do you need to pick blindly from the set $\{1, 2, \ldots, 2n\}$ if you want to be sure that one of the chosen numbers divides another chosen number?

E 9.17 Suppose 101 distinct numbers are written in a row (they do not need to be ordered by size). Prove that you can delete 90 of these numbers so that the remaining 11 numbers form either an increasing or a decreasing sequence. 3-4

E 9.18 Investigate the following number-theoretic variant of Problem 9.4: Let $n, N \in \mathbb{N}$. Find the smallest number $d \in \mathbb{N}_0$, depending on n and N, having the following property: for any natural number a at least one of the numbers $a, 2a, ..., Na$ lies within d of a multiple of n. 2-3

10 The extremal principle

Since the Form of the whole Universe
is most perfect, designed by a most wise Creator,
nothing at all takes place
that does not manifest, in some respect,
a Rule of Maximum or Minimum. [1]
(Leonhard Euler)

Extremes are fascinating. Who is the smallest, tallest, fastest, strongest? Everyday metaphors (the path of least resistance, to carry something to extremes etc.) show how deeply ingrained the idea of the extreme is in us. Moreover, the scientific view of the world reveals extremes everywhere: the soap bubble tries to minimise its surface area and is therefore spherical, chemical reactions strive towards a state of minimal energy, and so on.

The soap bubble shows us more: extreme forms often have particular properties, for example they may be very regular or symmetric. Reading this backwards we get a problem-solving strategy: If you are looking for an object with particular properties, try to characterise it by an extremal property. In this way extremes can be used to find special objects or at least to prove their existence. Looking out for extremes when you are solving problems can also be useful in many other ways. For example, it may help you find a starting point or structure your ideas.

In this chapter you will find examples for all of these different aspects of the extremal principle. Along the way you will learn about two fundamental inequalities and discover interesting facts about mirrors and billiards.

---

[1] Da aber die Gestalt des ganzen Universums höchst vollkommen ist, entworfen vom weisesten Schöpfer, so geschieht in der Welt nichts, ohne dass sich irgendwie eine Maximums- oder Minimumsregel zeigt.

D. Grieser, *Exploring Mathematics*, Springer Undergraduate Mathematics Series, https://doi.org/10.1007/978-3-319-90321-7_10

The extremal principle is an idea with far-reaching consequences. You will encounter it over and over again when you do mathematics, although it is rarely mentioned explicitly. Keep your eyes open, look out for it!

10.1 The general extremal principle

General extremal principle

Whenever something is extremal (largest, smallest etc.) particular structures will emerge.

For example, the shortest path between two points is a straight line (structure: no curves), the body of agiven volume with the smallest surface area is the ball (structure: perfect symmetry), and beautiful regular crystals appear when molecular configurations attain a state of minimal energy.

We now look at some examples in more detail.

Rectangles, squares of distances and an important inequality

? Problem 10.1

Among all rectangles with perimeter 20 cm, which one has the largest area?

Think about it!

Investigation and solution

▷ Denote the side lengths by a and b. Then the perimeter is $2a+2b$. From $2(a+b)=20$ we get $a+b=10$. The area is ab. So we are looking for a,b satisfying $a+b=10$ with largest ab. Let us check a few examples. Since a,b play exactly the same role in the problem, we may assume $a \leq b$. See Table 10.1.

a	b	ab
1	9	9
2	8	16
3	7	21
4	6	24
5	5	25

Table 10.1 Some rectangle areas

The table suggests that the area is largest if $a = b = 5$. However, the table is not a proof since a, b need not be integers. How can we prove it?

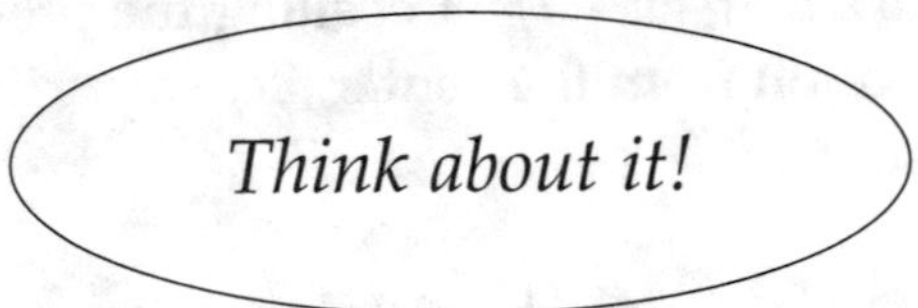

▷ **We focus on our goal**: the number 5 plays a special role. If a is less than 5 then b must be bigger by the same amount. So for any a, b with $a + b = 10$ we can write $a = 5 - x,\ b = 5 + x$ for some x. Then

$$ab = (5 - x)(5 + x) = 5^2 - x^2 = 25 - x^2\,.$$

Clearly this is less than or equal to 25 for all x, and equal to 25 if and only if $x = 0$, i.e. $a = b = 5$.

Therefore the rectangle of largest area is a square with sidelength 5 cm.

Review

The first thing you may have tried is eliminating b by writing $b = 10 - a$. Then you need to minimize $a(10 - a) = 10a - a^2$. It is not obvious how to proceed. One way would be to complete the square:

$$10a - a^2 = -(a^2 - 10a) = -(a^2 - 10a + 25 - 25) = -(a - 5)^2 + 25$$

and this yields the solution $a = 5$ as above. Another way would be to use calculus (if you know calculus).

Focussing on the expected answer 5 as in the solution above led us to the path of least resistance.

This is an example of the extremal principle: the extremal rectangle is a square, which is the most regular kind of rectangle.

? Problem 10.2

Which number between 0 and 2 minimizes the sum of the squares of its distances to 0 and to 2?

! Solution

If we call the number a then the two distances are a and $2-a$, so we want to minimize $a^2 + (1-a)^2$. Recalling the previous solution we look at the deviation from the center, i.e. we write $a = 1 - x$. Then $2 - a = 1 + x$, so

$$a^2 + (2-a)^2 = (1-x)^2 + (1+x)^2 = \\ 1 - 2x + x^2 + 1 + 2x + x^2 = 2 + 2x^2 .$$

Clearly this is minimal for $x = 0$, so for $a = 1$. !

Again the extremal principle is at work: the smallest sum of distances squared is attained in the symmetric case where a is halfway between 0 and 2.

A similar result holds for more than one point in the interval, see Exercise E 10.22. This is a simple one-dimensional model for a **crystal:** the sum of distances squared corresponds to the energy, and the state of minimal energy is the regular configuration.

The core of these two problems can be written as an inequality.

Theorem **(Inequality of the geometric, arithmetic and quadratic mean)**
Let $a, b \geq 0$. Then

$$\sqrt{ab} \leq \frac{a+b}{2} \leq \sqrt{\frac{a^2+b^2}{2}}.$$

Each inequality becomes an equality if and only if $a = b$.

We call $\sqrt{ab}$ the **geometric mean,** $\frac{a+b}{2}$ the **arithmetic mean** and $\sqrt{\frac{a^2+b^2}{2}}$ the **quadratic mean** of a and b. See Exercise E 10.4 for the significance of these means. There is a similar inequality for more than two numbers, see Exercise E 10.20.

Proof.
Let $c = \frac{a+b}{2}$ and $x = \frac{b-a}{2}$. Then $a = c - x$, $b = c + x$, so $ab = c^2 - x^2$ and $a^2 + b^2 = 2c^2 + 2x^2$, hence $ab \leq c^2 \leq \frac{a^2+b^2}{2}$. Now take square roots. Equality holds if and only if $x = 0$, i.e. $a = b$. q. e. d.

Make sure you understand why this inequality immediately yields the solutions of Problems 10.1 and 10.2.

Shortest paths

Everyone knows that the shortest path between two points is the straight line. Actually this is not easy to prove rigorously. But here we will take it for granted. A special case of this is:[2]

Theorem **(Triangle inequality)** In a triangle each side is shorter than the sum of the two other side lengths.

? Problem 10.3

Let a line g and two points A, B lying on the same side of g be given. How do you determine a point C on g for which the sum of distances from A to C and from C to B is minimal? See Figure 10.1.

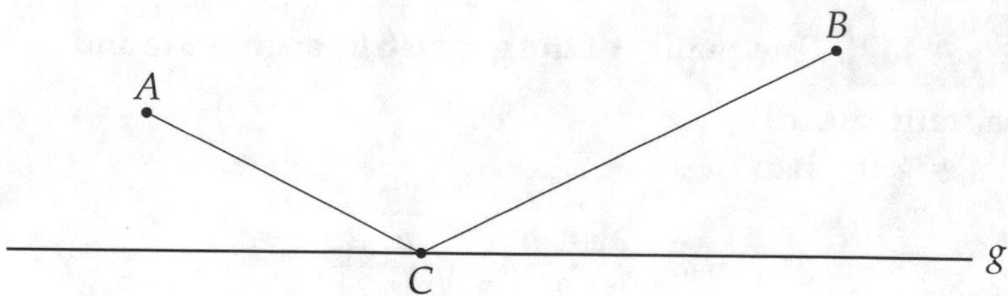

Figure 10.1 The problem of the shortest path touching the line g

Think about it. Then look at the following solution, which is beautiful and ingenious. Put it in your bag of tricks.

! Solution

Reflect B across the line g. Let B' be the mirror point. Draw the straight line from A to B'. It intersects g at a point C.
How do you proceed from here? Make a sketch.

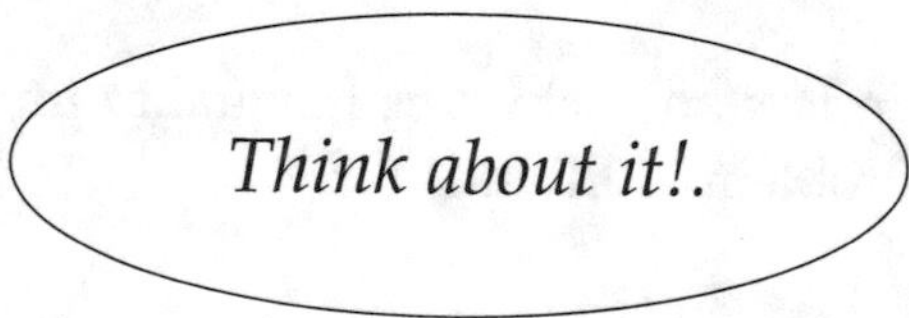

Let D be any other point on g. See Figure 10.2. on page 226

Since B and B' are mirror points they have the same distance to C. Therefore the distance from A to B via C is the same as the distance from A to B' via C.

Similarly, the distance from A to B via D is the same as the distance from A to B' via D.

By construction of C the path from A to B' via C is a straight line, and by the triangle inequality it is shorter than the path via D.

Therefore the path from A to B via C is shorter than the path via D. !

The solution has an interesting consequence:

[2] See Exercise E 10.8 for a direct proof of the triangle inequality.

Theorem Suppose the line g and the points A, B are as in Problem 10.3, and let C be the optimal point on g.

Then the angles that g makes with the lines CA and CB are equal: $\alpha = \beta$ in Figure 10.3.

Again this is an example of the general extremal principle.

Proof.
This follows immediately from the construction of C: The angles α and β' are opposite angles at the intersection of two lines and therefore equal, and $\beta' = \beta$ because B' is the reflection of B. Therefore $\alpha = \beta$.
q. e. d.

Remark

We call α the **angle of incidence** and β the **angle of reflection.** We have shown that the shortest path obeys the **law of reflection:** angle of incidence = angle of reflection. This is interesting for players of billiards: If you want to play a ball from A to B off the cushion g, try to imagine the reflection B' of B in g and aim at B'.
[3]

In Section 10.4 you will find more on this topic.

Other extremes

Extremal configurations need not always be symmetric: the rectangle of *smallest area* and perimeter 20 cm is a line segment 10 cm long (a 'degenerate' rectangle whose area is zero); the sum of squares of the distances from a to 0 and 2 is *largest* when $a = 0$ or $a = 2$. In both cases the extremum is attained at the boundary of the set of possibilities. This is also a type of particular property. So the general extremal principle still holds.

10.2 The extremal principle as problem solving strategy, I

By reading the general extremal principle backwards we obtain a problem-solving strategy:

[3]With real billiards you need to take into account that the balls are not points.

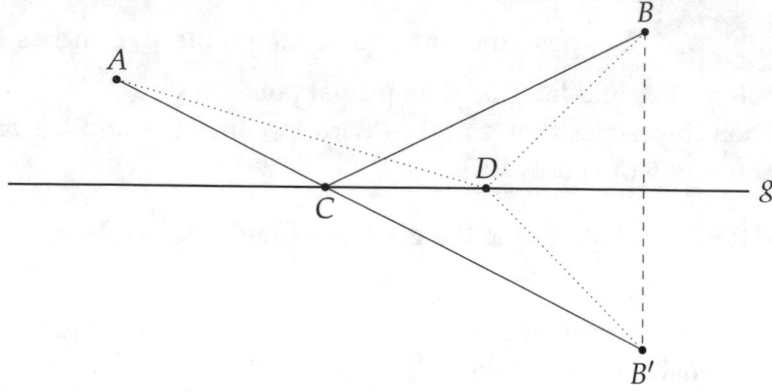

Figure 10.2 The solution

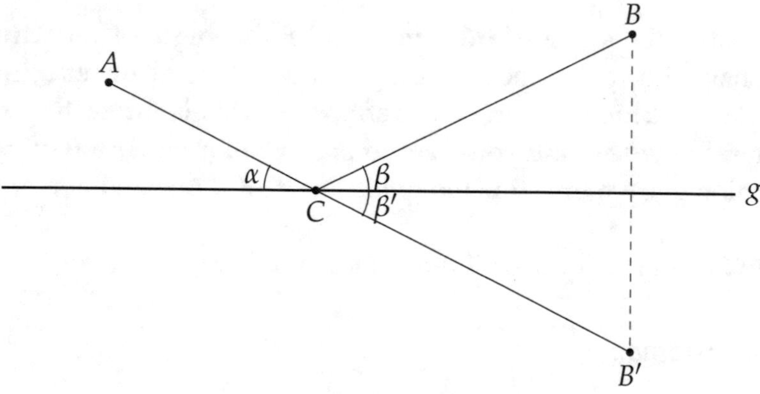

Figure 10.3 Equal angles

Extremal principle as problem-solving strategy

When looking for an object with particular properties, try to characterise it by an extremal property: largest, smallest, simplest, closest to having the desired property, etc.

Therefore you should reflect along a line which is on the table parallel to the cushion at distance one ball radius. Also you need to take into account that the ball will slightly dent the cushion, which has the effect that the angle of reflection is slightly bigger than the angle of incidence. This effect increases with ball speed.

So extrema can be useful for **proofs of existence**. In contrast to the pigeonhole principle (see Chapter 9) these proofs are usually **constructive,** as we will see.

We now consider an example, then read off the typical structure of such proofs, and then solve a more difficult problem using this idea.

A problem about tournaments

? Problem 10.4

In a tournament every player plays against every other player once. There are no draws. At the end every player makes a list containing the names of all players he defeated, and also of all the players defeated by a player whom he defeated.

Show that there is a player whose list contains the names of all other players.

Investigation and solution

▷ Look at a few examples to understand the problem. It is useful to **represent** the outcome of the games in the tournament **by a graph**: vertices are players, and edges correspond to games. Every edge is marked by an arrow, which points from the winner to the loser.[4]

See Figure 10.4 for an example. Player a defeated b and d, and b defeated c. Therefore a's list contains b, c and d. So player a satisfies the condition of the problem.

Player d's list only contains c and a, so d does not satisfy the condition. This is irrelevant since we only ask for *at least one* player with a complete list. Player b also has a complete list.

▷ How could we prove that there is a player with a complete list, *no matter* how each game ended? The difficulty is that we know almost nothing. In addition, the problem seems unapproachable because of the complicated description of the list.

▷ Could we replace the condition by a simpler one?

A player who has won many games seems likely to have a long list. So let us try this:

[4] A graph all of whose edges are marked by an arrow is called a **directed graph.**

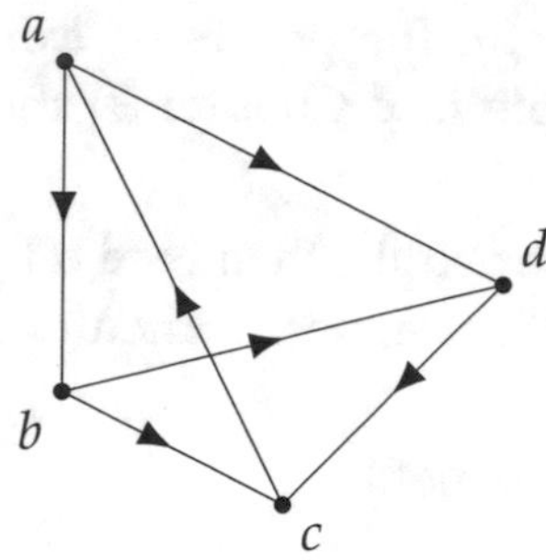

Figure 10.4 A tournament with 4 players

▷ *An attempt:* Let A be a player who has won the most games.[5] Can we conclude that A's list contains the names of all other players?

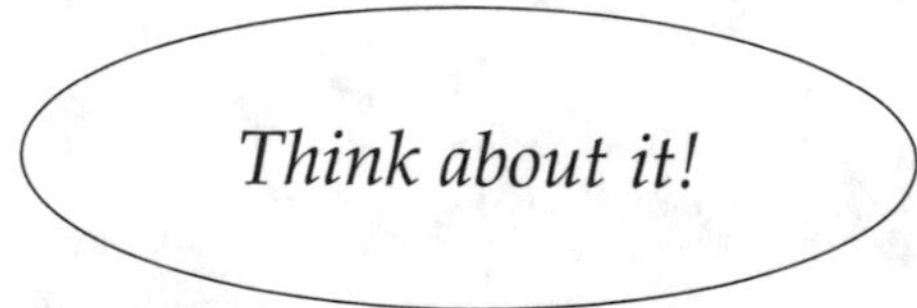

Suppose B is not on A's list. Then B must have defeated A. Also, B must have defeated all players that A has defeated, for if B had lost against one of them then B would be on A's list. So B has defeated all players that A has defeated, and also A. Therefore B has won more games than A. This is a contradiction to the assumption that A has won the most games.

Therefore the assumption that B is not an A's list must be wrong, so A's list contains the names of all other players.

How to use the extremal principle

The solution of Problem 10.4 shows how the extremal principle is typically used:

Problem: We are given a set M of objects and the specification of some property $\mathcal{E}$ that these objects can have.

[5] Why do we say 'a player', not 'the player'? Because 'the' would suggest that there is only one such player. But several players could have the same maximal number of victories, as in Figure 10.4, where both a and b have won twice.

We want a proof that there is an object in M having property $\mathcal{E}$.

Strategy: We look for a quantity whose maximisation allows us to infer property $\mathcal{E}$. Here a *quantity* is an assignment of a number to each object in M, that is, a function $q : M \to \mathbb{R}$.

Proof of existence: Suppose we have chosen a quantity q which we believe to serve our purpose.

1. Let $A \in M$ be an object for which $q(A)$ is maximal: $q(A) \geq q(B)$ for all $B \in M$.
2. We claim that A has property $\mathcal{E}$.
3. To prove this we assume that A does not have property $\mathcal{E}$. *Using this information we construct an object B satisfying $q(B) > q(A)$.* This is a contraction to the maximality of $q(A)$. Therefore A must have property $\mathcal{E}$.

Instead of maximising a quantity it may be more natural to minimise, depending on the problem.

Example

In Problem 10.4 M was the set of players, $\mathcal{E}$ was the property of having a complete list, and $q(A)$ was the number of games that A has won.

We are free to choose the quantity q: **If it works it is permitted.** The criterion is that step 3 can be done, in particular the construction of B. While we work on the problem the following may happen: we find a quantity q which we believe will work, but then we cannot do step 3. In this case we should look for another quantity q' – or try out different proof strategies, like induction or other ideas.

Looking for a suitable quantity q is a **creative process** and may require several attempts.

Once you have found a quantity q which works you get a **constructive procedure** to find an object with property $\mathcal{E}$: Start with any object A. If it has property $\mathcal{E}$ then we are done. If not, use step 3 to construct an object B with $q(B) > q(A)$. If B has property $\mathcal{E}$ then we are done, otherwise construct C with $q(C) > q(B)$ etc.

In the argument we use the following simple facts:

1. Every *finite* set of real numbers has a largest and a smallest element.

2. Every set of *natural numbers* has a smallest element.

A smallest element is also called a **minimum,** a largest element a **maximum.**

Remark

The second fact is also true for infinite sets of natural numbers, but not necessarily for infinite sets of positive *real* numbers. For example, the set $\{x \in \mathbb{R} : x > 0\}$ does not have a smallest element.
See Section 10.4 (after Problem 10.8) for more on the question of existence of extrema.

The problem of farms and wells

We now consider a problem where it is harder to find a quantity which we can maximise or minimise to solve the problem.

? Problem 10.5

We are given 2n points in the plane: n farms and n wells. We assume that no three of the points lie on a line. Is it possible to assign a well to each farm in such a way that each well belongs to only one farm and the straight paths from each farm to its well do not intersect?

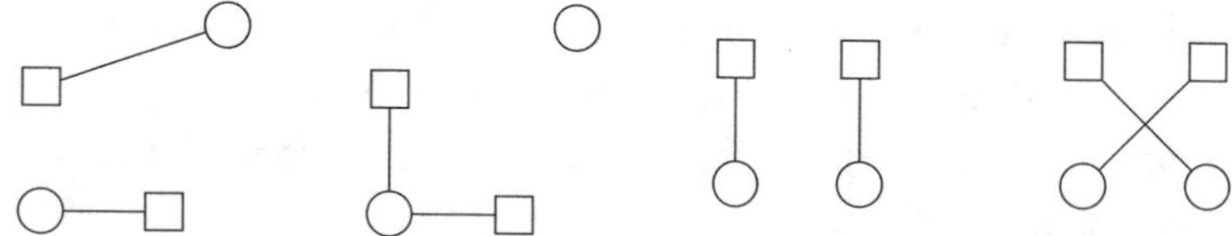

Figure 10.5 The first and third assignment of wells to farms are admissible, the others are not. Farms are squares and wells are circles. Assignments are indicated using line segments.

Investigation

▷ Let us call an assignment of wells to farms *admissible* if it satisfies the conditions, i.e. if it is *bijective* and *intersection-free*. Figure 10.5 shows a few examples with $n = 2$. These are deceptively simple. For large n things could look quite complicated. If you try a few more examples you will notice that an admissible assignment seems to always exist. So we formulate:

Conjecture: There is always an admissible assignment.

▷ How could we prove this? The difficulty is that there are so many possibilities for the positions of farms and wells. As a first idea we could put ourselves in the position of a farmer:

First attempt: To each farm assign the well which is closest.

If the farms lie as in the first picture in Figure 10.5 then both farms get the same well (second picture), so the assignment is not bijective.

So that didn't work: it was too short-sighted to take the individual points of view of each farmer as a basis. We should keep in mind the total configuration.

▷ Let us try to use the extremal principle. Let us **make a plan**. What are we looking for? Among all bijective assignments we are looking for one without intersections. It is natural to try the following.

Second attempt: Consider a bijective assignment with the minimal number of intersections.[6]

This seems promising since for a solution of the problem this number must be zero, hence minimal.

▷ We want to prove: If A is a bijective assignment with minimal number of intersections, then A has no intersection. To prove this, we assume that A has an intersection. Then there would be two farms and wells with assignments as in Figure 10.6 (solid lines; there could be more farms and wells which are not drawn). We want to change this to a different bijective assignment B having fewer intersections than A. How could we do this?

[6] So in the general scheme of the previous section M is the set of bijective assignments, $\mathcal{E}$ is the property of having no intersections, and $q(A) =$ number of intersections in $A \in M$.

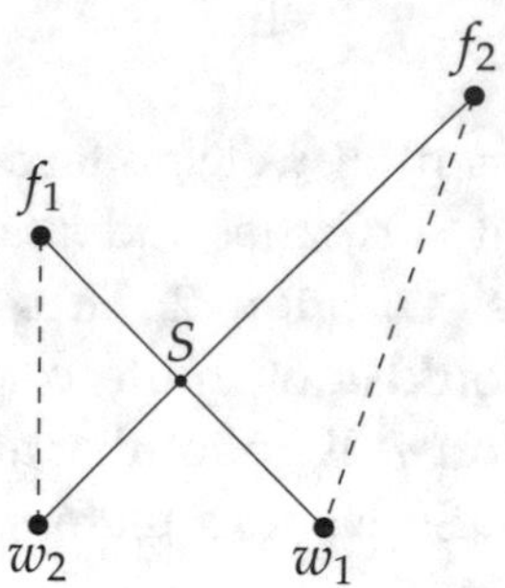

Figure 10.6 A simple idea for reducing the number of intersections. Here farms and wells are represented by dots.

▷ We want to remove the intersection S. That's easy: replace the solid lines by the dashed lines and leave all other lines (i.e. pairs farm – well) unchanged. Let us call the new assignment B. Does B necessarily have fewer intersections than A?

▷ The left-hand picture in Figure 10.7 shows that this is not necessarily true. In this example, while removing the intersection S (using the dashed lines) we produce a new intersection T. So B has the same number of intersections as A has. Even worse, you can easily modify the example so that B has more intersections than A.

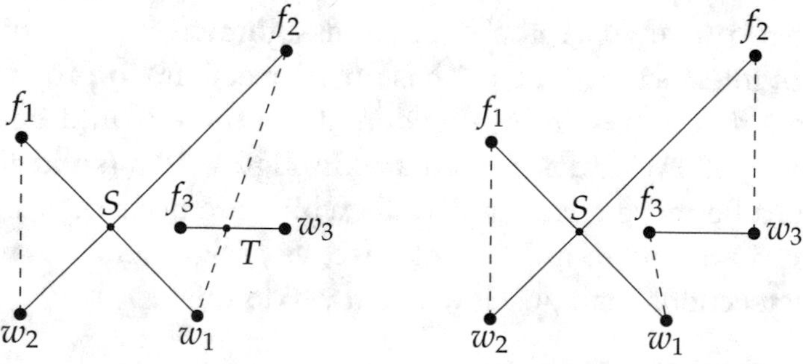

Figure 10.7 A problem with the second attempt and a solution for this example

▷ We see that our way to construct B from A was too simple-minded. The right-hand picture in Figure 10.7 shows a way to get an intersection-free assignment in this case (dashed lines). **Can we find a general construction,** which in this example replaces the solid lines (assignment A) by the dashed lines as in the right-hand picture (assignment B), and which reduces the number of intersections for *any* A?

For example, the construction must also work if there are more farms in the triangle Sw_1f_2 (or in the triangle Sw_2f_1) and A assigns them to wells outside the triangle.

▷ The complexity of the situation increases dramatically. It is not at all clear how to proceed.[7]

So let us try something completely different. In the second attempt we minimised the number of intersections; that didn't work. Are there other quantities whose minimisation might work?

We could try to combine elements of the first and second attempt: we consider the distances from the farms to the wells – however, this time not individually, but with a view to the entire configuration. So we should cook up a single overall number from all of these numbers.

The simplest thing is to add all these numbers. Surprisingly this works.

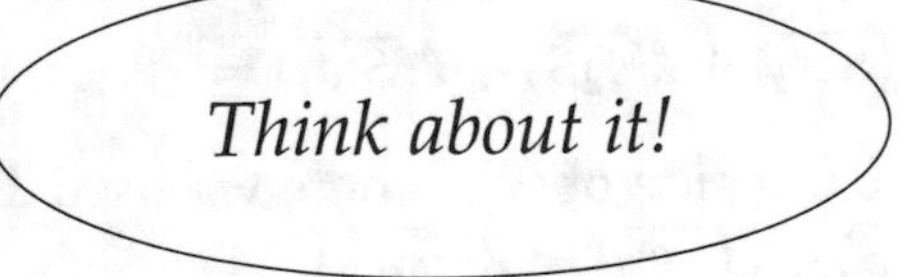

! Solution of Problem 10.5

There always is such an assignment.

Proof: Among all bijective assignments of wells to farms consider one which has the *smallest sum of all distances from the farms to their wells.* Call this assignment A. We show that A has no intersections and thus solves the problem.

[7] If you do find a way to finish this attempt successfully, please let me know – D.G.

Suppose this was not the case. Then there would be wells w_1, w_2 assigned to farms f_1, f_2, respectively, so that the line segment $\overline{f_1w_1}$ intersects the line segment $\overline{f_2w_2}$. Let S be the point of intersection. So we have a situation as in Figure 10.6, since no three points lie on a line. We didn't draw the other farms and wells.

Claim: Let B assign the well w_2 to the farm f_1 and the well w_1 to the farm f_2 (dashed lines) and leave the assignments of all other wells to their farms as in A. Then B has a smaller sum of distances than A.

This would be a contradiction to the minimality of A. Therefore A is admissible.

It remains to prove the claim. Since all other assignments stay the same, it suffices to show that the two dashed lines together are shorter than the two solid lines together:

$$\ell(\overline{f_1w_2}) + \ell(\overline{f_2w_1}) \stackrel{!}{<} \ell(\overline{f_1w_1}) + \ell(\overline{f_2w_2}) \qquad (*)$$

where $\ell(\overline{PQ})$ denotes the length of the line segment from a point P to a point Q. To prove $(*)$ we use the triangle inequality, applied to the triangles w_2Sf_1 and w_1f_2S:

$$\ell(\overline{f_1w_2}) < \ell(\overline{f_1S}) + \ell(\overline{Sw_2}), \quad \ell(\overline{f_2w_1}) < \ell(\overline{f_2S}) + \ell(\overline{Sw_1}).$$

Adding these inequalities we obtain the left hand side of $(*)$ on the left, and on the right

$$\ell(\overline{f_1S}) + \ell(\overline{Sw_2}) + \ell(\overline{f_2S}) + \ell(\overline{Sw_1}) = \ell(\overline{f_1w_1}) + \ell(\overline{f_2w_2}),$$

that is, the right hand side of $(*)$, where we used $\ell(\overline{f_1S}) + \ell(\overline{Sw_1}) = \ell(\overline{f_1w_1})$ and $\ell(\overline{f_2S}) + \ell(\overline{Sw_2}) = \ell(\overline{f_2w_2})$. !

Review of Problem 10.5

- ☐ The idea of minimising the natural quantity $q(A) =$ (number of intersections for the assignment A) did not lead to a proof. This may seem paradoxical since we know that this number *is* minimal for the assignment we are looking for. However, at this point in the argument we do not know yet that there is an intersection-free assignment; this is only a conjecture. It could have turned out that the conjecture was wrong, i.e. that there is a configuration of farms and wells for which every bijective assignment has intersections.

- It is essential that the construction of B in the second step of the proof works for *all* A that have intersections. Only in this way we get a complete proof. Examples can help us to get an idea, but they are not sufficient.
  For $q(A) =$ (number of intersections for the assignment A) we did not succeed in finding such a construction.
- However, the choice $q(A) =$ (sum of distances for A) quickly led to success. Note that in the construction of B with smaller distance sum we could have created new intersections; but this is irrelevant since only the sum of distances matters for the argument. Make sure that you understand this important point. See also the algorithmic point of view below.
- Quite generally in geometric problems it makes sense to look at maxima or minima of *geometric* quantities, i.e. those related to lengths, distances, areas etc. Of course this is only a heuristic and not a universal recipe.

Let us reconsider the problem of farms and wells from an **algorithmic point of view:** we want to give an algorithm for finding an admissible assignment. We try the following simple algorithm:

1. Choose any bijective assignment. If it has no intersections we are done. Otherwise proceed to step 2.

2. Choose an intersection and remove it by changing the assignments for the two farms involved as in Figure 10.6.

3. Repeat step 2 until you get an assignment without intersections.

Does this algorithm terminate at some point? This is not obvious at all. We saw that when removing one intersection we might create several others, so it seems possible that we might get into a loop, arriving at an assignment that we have already considered some steps before. Then we could repeat the loop forever.

However, our solution of Problem 10.5 implies that the algorithm does terminate. How?

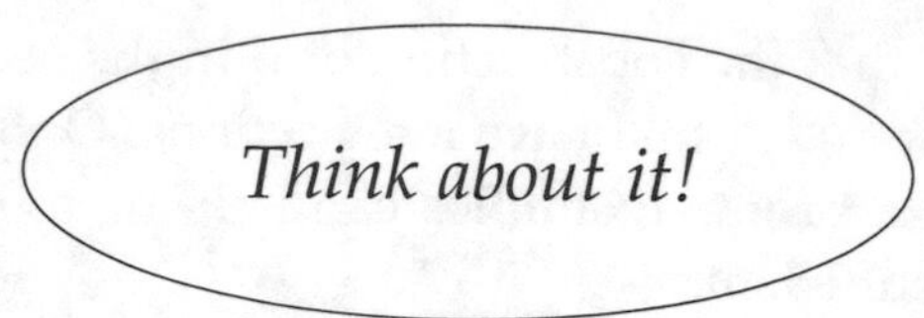

We showed that the sum of distances of farms to their assigned wells decreases every time that we do step 2 of the algorithm. This already shows that no loops are possible. More to the point, there are only finitely many numbers that appear as sums of distances since there are only finitely many assignments. Therefore we must arrive at the smallest of these numbers after a finite number of steps (unless the algorithm stopped before that). Then our proof shows that the corresponding assignment has no intersections, so the algorithm stops.

Summarizing, we have seen that although the number of intersections may increase while we run the algorithm, at the end there will be no intersections.

10.3 The extremal principle as problem solving strategy, II

We saw that the extremal principle can be a tool for proofs of existence. But considering extremes can also be useful as a more general idea.

Extremal principle as problem solving strategy, II

In a complex situation look at objects with an extremal property.

In the simplest case, looking at extremes may help you to formulate an idea: we saw examples of this in the proofs of the theorems on division with remainder in Section 8.1 and on rational approximation in Section 9.3.

We now consider some examples where the general idea of extremes helps to solve problems.

A problem about averages

? Problem 10.6

1000 numbers are written along the perimeter of a circle. Each number is the average of its two neighbours. Show that all numbers are equal.

Think about it!

Investigation and solution

▷ *First solution:* Look at one of the numbers. Since it is the average of its neighbours, there are two possibilities:

1. It is equal to its neighbours, or

2. one neighbour is larger, the other smaller.

In the second case consider the larger neighbour. By the same reasoning its other neighbour must be still larger. Continuing, we see that the numbers must get larger and larger while we go around the circle in one direction. This leads to a contradiction once we arrive at the starting number since it cannot be bigger than itself.

Therefore the first case must apply. Since the number we started with was arbitrary, we have proved that each number is equal to its neighbours. Therefore all numbers must be equal.

Here is another argument which is shorter:

▷ *Second solution:* Let x be the *smallest* of the 1000 numbers. Since it is the average of its neighbours, these neighbours must equal x since otherwise one of them would be smaller than x, contradicting minimality of x. So both neighbours are equal to x, hence also minimal. By the same argument, their neighbours must also equal x, then the next neighbours etc. Since in this way you reach all numbers, they must all equal x.

The two solutions are very similar; it almost seems as if the second is a rephrasing of the first. This is not so: the second solution, using the extremal principle, can be generalised to a two-dimensional version of the problem, but the first can't. See Exercise E 10.14.

Infinite descent

Infinite descent is a type of argument which is closely related to the extremal principle. Here is an example.

? Problem 10.7

Can three times a square number equal the sum of two square numbers?

Investigation

▷ The question is: Are there natural numbers a, b, c satisfying

$$3a^2 = b^2 + c^2\,?$$

Try to find a solution by testing a few numbers.

▷ Have you seen a **similar problem** before?

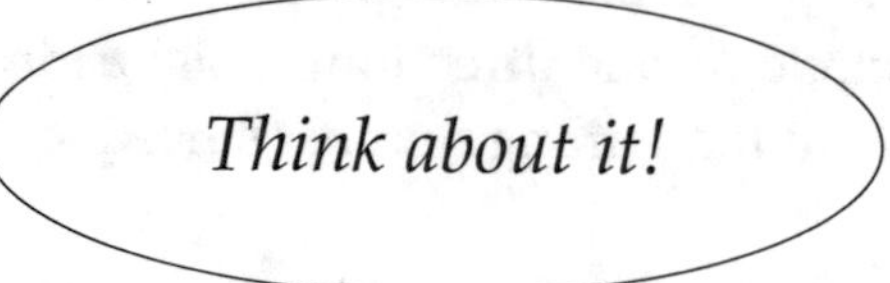

▷ Maybe the problem reminds you of the proof that $\sqrt{2}$ is irrational, see Section 7.2. There we had to prove that the equation $2q^2 = p^2$ has no solution in natural numbers.

We don't know yet whether our problem has a solution. If we could argue as we did for $\sqrt{2}$ then we might be able to prove that there is no solution. Try it!

▷ How did we proceed with the equation $2q^2 = p^2$? We first showed that p is even, using that $2q^2$ and hence p^2 is even.

In our problem we have a 3, not a 2, so we might try to argue with divisibility by 3, not 2.

▷ So let us suppose we had a solution of $3a^2 = b^2 + c^2$. Then $b^2 + c^2$ is divisible by 3. Unfortunately we cannot conclude from this directly that b or c are divisible by 3. How to go on?

▷ The generalisation of even/odd is the remainders modulo 3, see Chapter 8. Let us check the possible remainders modulo 3 of b and

b^2. There are three cases:

$$\begin{aligned} b \equiv 0 \mod 3 &\Rightarrow b^2 \equiv 0^2 = 0 \mod 3 \\ b \equiv 1 \mod 3 &\Rightarrow b^2 \equiv 1^2 = 1 \mod 3 \\ b \equiv 2 \mod 3 &\Rightarrow b^2 \equiv 2^2 = 4 \equiv 1 \mod 3 \end{aligned}$$

We see that b^2 can only have the remainder 0 or 1 (compare Exercise E 8.10). Can you go on from here?

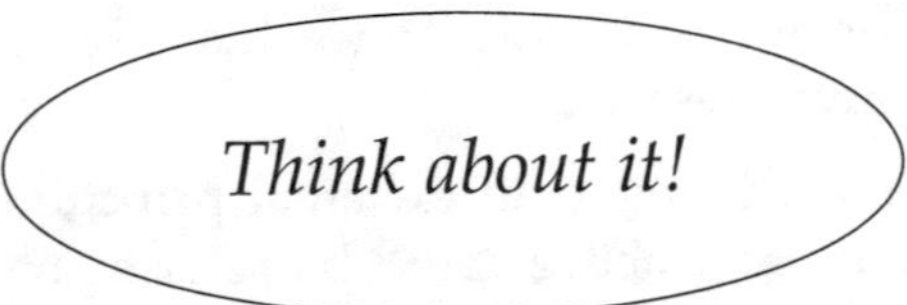

▷ For the same reason c^2 can only have the remainder 0 or 1, so $b^2 + c^2$ can only have the remainder 0, 1 or 2, and the remainder 0 can only occur if both b^2 and c^2 have remainder 0.

▷ We summarise: From $3a^2 = b^2 + c^2$ we can conclude that $b^2 + c^2 \equiv 0 \mod 3$ and from this that $b \equiv c \equiv 0 \mod 3$. Therefore we can write $b = 3B$, $c = 3C$ with integers B, C.

▷ When we plug these in, we get $3a^2 = 9B^2 + 9C^2$, so $a^2 = 3B^2 + 3C^2 = 3(B^2 + C^2)$. Therefore, $a^2 \equiv 0 \mod 3$, and by the cases above we conclude that $a \equiv 0 \mod 3$. Then we can write $a = 3A$, and plugging this in yields $9A^2 = 3(B^2 + C^2)$, hence

$$3A^2 = B^2 + C^2 .$$

▷ How can we use this? This is the same equation as the one we started with. But A is smaller than a (since $a = 3A$), and similarly for B, C. This easily leads to a contradiction, see below.

! Solution of Problem 10.7

There are no such square numbers.
Proof: Suppose there were natural numbers a, b, c satisfying $3a^2 = b^2 + c^2$. Then $b^2 + c^2$ is divisible by 3, and the argument above shows that then b and c are divisible by 3, so $b = 3B$, $c = 3C$ with $B, C \in \mathbb{N}$.

Again by the argument above it follows that a is divisible by 3, so $a = 3A$ with $A \in \mathbb{N}$. We obtain $3A^2 = B^2 + C^2$.

We have shown: If $3a^2 = b^2 + c^2$ then a, b, c are divisible by 3, and $A = \frac{a}{3}, B = \frac{b}{3}, C = \frac{c}{3}$ are solutions of the same equation. Then by the same argument A, B, C must be divisible by 3 etc. In this way we would obtain an infinite sequence of ever smaller and smaller solutions, all natural numbers. This is clearly impossible. !

> *Proof analysis*: This was an **indirect proof.** The core of the argument was to consider the remainders modulo 3 of square numbers. The method by which we arrived at a contradiction is called **infinite descent.**
>
> Infinite descent is really an **extremal principle** argument in disguise; the proof above could be reformulated as follows: Suppose there is a solution a, b, c. Let a_0 be the smallest value of a among all solutions. Construct A, B, C as before. This is a solution again, and because $A = \frac{a_0}{3} < a_0$ we get a contradiction to the fact that a_0 was minimal. Therefore a solution cannot exist.

10.4 Going further: optimisation, mirrors and billiards

We encounter extrema (that is, maxima and minima) in mathematics in two types of contexts:

1. Determining extrema as a goal in itself: often motivated by applications, we want to determine how to make something fastest, cheapest, biggest, smallest etc. This is called **optimisation.**

2. Using extrema as a means for other purposes, for example for proofs of existence. You saw some examples in this chapter.

In this section you will find more examples and a more extensive discussion of shortest paths. Even if you don't understand all of the words in the explanations that follow, keep reading anyway. You will be astonished how universal the extremal principle is.

In the books (Nahin, 2004) and (Hildebrandt and Tromba, 1996) you will find more information about many of these topics and their history.

Optimisation problems

Here are some examples of optimisation problems from different strands of life and mathematics.

- An engineer who constructs a bridge wants it to be as stable as possible, while at the same time it should be light-weight and cheap.[8]
- A company wants to use its resources (personnel, means of production) in such way that the profit is maximised.
- Given several points in the plane, find a curve which is 'simple' and comes as close as possible to the points. A typical version of this problem is the least squares problem of **linear regression** in statistics. Here real numbers $x_1 < x_2 < \cdots < x_n$ and $y_1, y_2, \ldots, y_n$ are given, and we look for those numbers a, b for which the linear function $f(x) = ax + b$ produces values at the x_i which are *closest in mean square* to the y_i, i.e. for which $(y_1 - f(x_1))^2 + \cdots + (y_n - f(x_n))^2$ is minimal.
- The **isoperimetric problem** is the question which planar figure with given area has the smallest perimeter, or which body with given volume has the smallest surface area. The answers are the circle and the ball.[9] It is not easy to prove this rigorously. A related problem is to find the surface of smallest area whose boundary is a given closed curve in space.[10]
- Suppose A, B are points in space, where A lies higher than B and not vertically above it. We want to build a track on which a ball can roll from A to B. What shape must the track have so that the time needed is as short as possible? Maybe your first guess is that the track should be as short as possible, hence a straight line. However, it turns out that the total time will be shorter if the track is very steep at the beginning and then gets flatter – and maybe even turns upwards near the end, for certain positions of A and B!

---

[8] Not to mention aesthetic considerations.

[9] An equivalent problem is to find the planar figure with given perimeter and largest area, and similarly for bodies. Compare Problem 10.1.

[10] Imagine the curve is a bent wire. Put the wire into soapy water. When you pull it out carefully you will see a soap film, and this will be a surface of minimal area.

In this way the ball will pick up more speed in the beginning and arrive sooner than for the straight track, although its path is longer. What is the precise shape of the optimal track? To answer this you need to know differential equations. The best curve is called the **brachistochrone.** This sort of optimisation problem, where we are looking for a curve, not a number, is called a **variational problem.**

- The **travelling salesman problem:** You want to visit n cities. The cities are fixed in advance, but you can choose the order in which you visit them. How should you travel in order to minimise the length of the journey?[11]

- The problem of **optimal mass transport:** You have a pile of sand and a hole, and you want to put the sand in the hole. The hole is big enough for all the sand. Moving the sand takes work. The problem is: where should each grain of sand go in order to minimise the total work? What makes this problem non-trivial is that the hole can have a different shape from the pile.

- The problem of **densest sphere packing:** how do you pack a large number of spheres of equal size so that the gaps are as small (in total) as possible? If you try stacking oranges you will quickly arrive at a guess for the best packing. KEPLER conjectured in 1611 that this packing is optimal. However, to *prove* that this is the optimal packing is extremely difficult and was done only in 1998.

Examples of the extremal principle in different areas of mathematics

The extremal principle as a method of proof is used in all areas of mathematics. You have already encountered a few instances, for example from number theory. Here are a few more examples. Only a few keywords are given, and the areas of mathematics are indicated in parentheses. You will learn about some of these as an undergraduate student of mathematics; for some of them you may have to wait until postgraduate studies.

- The supremum axiom for the real numbers (analysis).

[11]There is no known algorithm which solves this problem generally for $n = 100$ in an acceptable time (the age of the universe, say)!

- Proving the mean value theorem using the extreme value theorem (analysis).

- Proving that symmetric matrices A have eigenvalues, by minimising or maximising the associated quadratic form $f(x) = \langle x, Ax \rangle$ over the unit sphere (linear algebra, functional analysis, partial differential equations).

- ZORN's lemma, which can be used to prove that every vector space has a basis, or to prove the HAHN-BANACH theorem and many other important theorems (linear algebra, functional analysis, etc.).

- The first complete proof of the fundamental theorem of algebra (every polynomial has a zero in the complex numbers), published by ARGAND in 1814, uses the extremal principle (complex analysis).

- The maximum principle for harmonic functions, which is useful for analysing these functions and also for proving that the DIRICHLET problem can have at most one solution (analysis, partial differential equations).

- Proving that in a HILBERT space the projection to a given closed subspace exists, via minimising distances from points to the subspace; this is the core for the existence theory for solutions of partial differential equations using the HILBERT space method (functional analysis, partial differential equations).

- In the variational calculus the extremal principle is used systematically for solving differential equations.

- Investigation of the topology of a space via MORSE theory (topology).

Keep your eyes open; you will find many more examples!

Another beautiful example is the **rainbow:** Why are there rainbows, why are they circular arcs, and how can you calculate their position? This also amounts to solving an extremal problem (see (Nahin, 2004)).

Derivatives: a new look at shortest paths, light rays, billiards etc.

Calculus provides an important technique for solving many optimisation problems: to find the extrema of a function you find the zeroes of its derivative.[12] This is the general idea, but there are some details to watch out for: an extremum must be a stationary point only if it is attained at an *interior point* of the domain of definition; an interior, stationary point need not be a (local) extremum – it could also be a point of inflection. So you need to check boundary points separately, and for each interior, stationary point you need to check whether it is a local extremum. Finally, if you are only interested in the global extrema, then you have to check for that. But in most cases these details are easy to deal with.

Using this technique you can solve Problems 10.1 and 10.2 and give a proof for the inequality of the geometric, arithmetic and quadratic mean. However, the elementary solutions given in this book are more appropriate considering the simplicity of these problems.

The relationship of extrema and derivatives gives another hint why the general extremal principle holds: if the function $x \mapsto f(x)$ has an extremum x_{extr} in the interior of its domain of definition then $f'(x_{\text{extr}}) = 0$. So the point x_{extr} is in the solution set of this equation, and the solution sets of equations often have particular structure.

As an example we give a calculus solution of Problem 10.3: with notation as in Figure 10.8 (where d is the distance from P to Q) we want to find the point C, i.e. the number x, so that $f(x) = \ell(\overline{AC}) + \ell(\overline{CB})$ is minimal.[13] By PYTHAGORAS' theorem we have $f(x) = \sqrt{a^2 + x^2} + \sqrt{b^2 + (d-x)^2}$, and after a short calculation you get the derivative

$$f'(x) = \cos\alpha - \cos\beta. \tag{10.1}$$

At a minimum $x_{\min}$ of f the equation $f'(x_{\min}) = 0$ must hold (since the domain of definition is $\mathbb{R}$, so all points are interior), hence $\cos\alpha = \cos\beta$, which implies $\alpha = \beta$ since $\alpha, \beta \in (0, \pi)$.

This is a new proof of the theorem, stated after Problem 10.3, that the shortest path from A to B obeys the reflection law. This new

[12] These are also called the *stationary points* of the function. Here we always assume that the function is differentiable at all relevant places.

[13] $\ell(\overline{AC})$ denotes the distance between the points A and C.

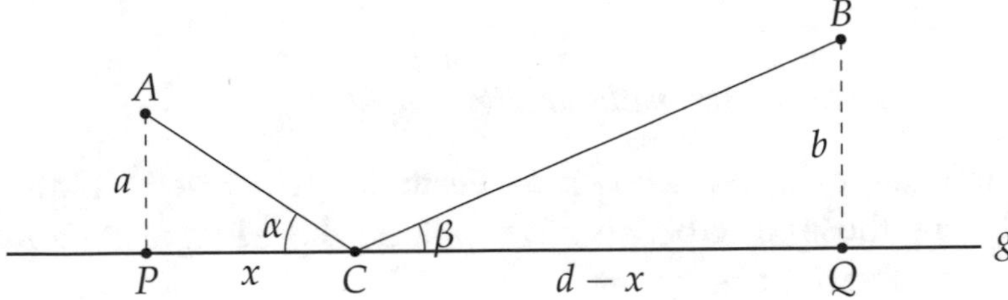

Figure 10.8 Notation for the calculus solution of Problem 10.3; C is not the minimum in the picture

proof does not use the mirror point B' (compare Figure 10.3), and therefore it has the advantage of being applicable in situations where the mirror construction doesn't work. For example, we can replace the line g by a curve. Then one can prove in a similar way:

Theorem Let c be a curve and let A, B be two points lying on the same side of c, see Figure 10.9. Let C be an interior point on c for which the sum of the distances $\ell(\overline{AC}) + \ell(\overline{BC})$ is extremal. Then $\alpha = \beta$, that is, angle of incidence $=$ angle of reflection at C.

Conversely, at any point C where $\alpha = \beta$ the function $f(C) = \ell(\overline{AC}) + \ell(\overline{BC})$ is **stationary** at C, i.e. it has a local maximum or minimum or an inflection point.

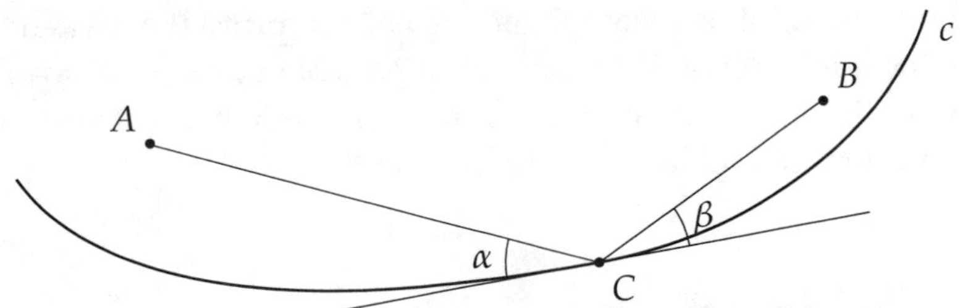

Figure 10.9 Reflection in a curved mirror; α, β are the angles with the tangent line of the curve c in the point C

If we think of c as a mirror then $\alpha = \beta$ means that a light ray coming from A gets reflected to B. So we can rephrase the theorem as follows:

Longest and shortest paths are light paths.

The converse of this is known as **Fermat's principle:**[14] Light will travel along those paths between two points whose length is stationary under small variations.[15]

Looking at the previous theorem from a different angle we get another interesting consequence. Consider the following problem.

? Problem 10.8

Let c be a curve and let A, B be two points lying on the same side of c. Does it follow that there is a point C for which the angles α and β are equal?

So we want to send a billiard ball from A to B via the cushion c. Is that possible?

! Solution

(incomplete) Choose an interior point C on c so that $\ell(\overline{AC}) + \ell(\overline{CB})$ is minimal. Then the theorem above implies $\alpha = \beta$. So a point C as desired exists. !

This is a beautiful example of the extremal principle as a tool for proofs of existence. See Exercise E 10.25 for an interesting variant of this problem.

Why is the solution incomplete? We need to prove the existence of C, i.e. the minimum of f! For the set of possible function values $f(C)$ is neither finite nor a set of natural numbers, so it is not obvious that it has a minimum. How can we complete the proof?

---

[14]Pierre de FERMAT, 17th century. The principle was generalised in the 18th and 19th century by EULER, LAGRANGE, MAUPERTUIS and HAMILTON to the 'principle of least action', which aims to describe all physical processes in a similar way.

[15]This is the version for homogeneous media, where the speed of light is constant. If you replace the length of the path by the time of travel then this still holds for inhomogeneous media. If you apply this to the transition between two media then you get the law of refraction (SNELL's law).

1. The *extreme value theorem* of analysis says that any continuous function on a bounded closed interval has a minimum and a maximum. By parametrising the curve we can think of f as a function on an interval. It is bounded and closed if we include the endpoints of c. Continuity of f is easy to show using the triangle inequality. So we know that f does have a minimum.

2. In order to use the theorem above we need to show that the minimum is attained in the interior of the curve. This is not clear, and indeed need not be true: if c is an interval on the x axis and A, B both lie above and to the left of c then the minimum is attained at a boundary point, and clearly there is no reflection point C on c. So the answer to the problem is yes only if it can be guaranteed that the minimum is attained at an interior point.

   This is the case, for example, if c is a closed curve, such as an ellipse.

In many situations where we would like to use the extremal principle the function f depends on several variables. The extreme value theorem is then still true if each variable can vary independently on a bounded closed interval. However, this is an unnatural restriction on the domain of definition of f. Here it is more natural to use the notion of a **compact set.** You can learn more about this central notion of mathematics in books on analysis or topology.

10.5 Toolbox

Starting from the general extremal principle, a general scientific idea, we arrived at the extremal principle as a problem-solving strategy. We first used this strategy as a means for proofs of existence: **try to characterise the desired object by an extremal property.** Often it is not obvious which property will work for this, and a **dedicated search** and several attempts may be necessary. There are **no limits to your creativity** (as long as you argue correctly, of course). In a more general sense, the extremal principle as a strategy tells you that in any kind of complex situation it is worth looking for extremes. One variant of this idea is the method of **infinite descent**, which sometimes helps to prove that an equation has no solution in natural numbers.

Exercises

1 **E 10.1** Choose a natural number bigger than two. Take its square root and round it down, then take the square root again and round it down etc. Show that at some point you will obtain 2 or 3. How is this related to the extremal principle?

1-2 **E 10.2** Suppose the product of two positive real numbers is 20. What is the smallest possible value for their sum? What is the largest?

1-2 **E 10.3** In Problem 10.2 replace the squares of the distances by the reciprocals of the distances and solve the problem.

2 **E 10.4** Let $0 < a \leq b$ be given.

a) A car moves at speed a (kilometers per hour) for an hour and then at speed b for an hour. What is its average speed? (average speed = total distance / total time)

b) A car moves at speed a for 100 km and then at speed b for another 100 km. What is its average speed?

c) Prove the **inequality of the harmonic and the geometric mean,** $\frac{2}{\frac{1}{a}+\frac{1}{b}} \leq \sqrt{ab}$. Deduce the inequality $\frac{2}{\frac{1}{a}+\frac{1}{b}} \leq \frac{a+b}{2}$. Also argue without calculation why the latter inequality holds, using a) and b).

d) Find a real-life context where the geometric mean $\sqrt{ab}$ appears.

2 **E 10.5** In Problem 10.4 we looked at a player who has won the most games, and this led to a solution. Instead we could have looked at a player with the longest list. Can you find a solution of the problem from here?

1-2 **E 10.6** Does the argument in the solution of Problem 10.5 also work if you use $q(A) =$ (the biggest distance from a farm to a well) instead of the sum of distances?

2 **E 10.7** One of the fundamental facts of analysis is that $\mathbb{Q}$ is a **dense** subset of $\mathbb{R}$. This means: If $a < b$ are real numbers then there is a rational number r satisfying $a \leq r \leq b$. Prove this.

2 **E 10.8** Prove the triangle inequality (stated before Problem 10.3) using coordinates in the plane, for example as follows: We may

assume that one vertex of the triangle is the origin and one vertex lies on the positive x axis, so $A = (0,0)$, $B = (b,0)$ where $b > 0$. Let the third vertex be $C = (x,y), y \neq 0$. Then the side lengths are $\ell(\overline{AB}) = b$, $\ell(\overline{AC}) = \sqrt{x^2+y^2}$ and $\ell(\overline{BC}) = \sqrt{(x-b)^2+y^2}$ by PYTHAGORAS' Theorem. Prove the inequality $\sqrt{(x-b)^2+y^2} < \sqrt{x^2+y^2}+b$.

E 10.9 Prove that every triangulation of a convex n-gon (see Problem 2.4) has a boundary triangle, i.e. a triangle whose vertices are successive vertices of the n-gon. 2-3

E 10.10 Can five times a square be the sum of two squares? How about seven times a square? 2

E 10.11 Prove that one cannot find four natural numbers a, b, α, β satisfying the equation $a^2 + b^2 = 3(\alpha^2 + \beta^2)$. 2

E 10.12 Show that the equation $4x^4 + 2y^4 = z^4$ has no integer solution except $x = y = z = 0$. 2-3

E 10.13 Is there a square number which lies in the middle between two other square numbers? 2

E 10.14 2

a) Suppose at each integral point $n \in \mathbb{Z}$ of the number line a natural number is written. Also suppose that each number is the arithmetic mean of its two neighbours. What can you conclude about the numbers?

b) Suppose at each lattice point of the plane (i.e. at each point with integer coordinates) a natural number is written. Also suppose each number is the arithmetic mean of its four neighbours. What can you conclude about the numbers?

c) What can you conclude if instead of a natural number you write a positive real number at each n in a)?

E 10.15 Let a graph be given. A black and white colouring of its vertices is called *desegregated* if each white vertex has at least as many black neighbours as white neighbours, and vice versa. 2-3

Prove that there is a desegregated colouring of the vertices.

3 E 10.16 Suppose in a group of people everyone knows at most three other people. Prove that it is possible to divide the group into two smaller groups such that in each of the small groups every person knows at most one other person in the same small group.

2-3 E 10.17 Consider a convex polygon and an arbitrary point P in its interior. Each side can be extended to an infinite line, and we can drop a perpendicular from P to any of these extended sides. Show that there must be at least one case where the perpendicular hits the side itself in its interior. See Figure 10.10.

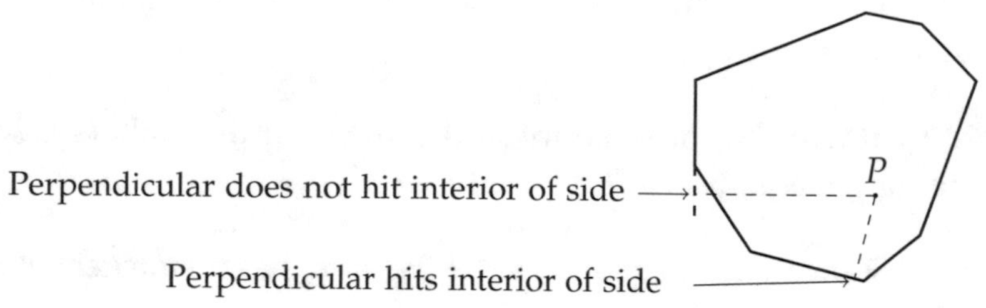

Figure 10.10 Exercise E 10.17

3-4 E 10.18 Several barrels of petrol are arranged along a circular track. The barrels may be filled to different levels, and may be positioned at irregular intervals. Suppose the total amount of petrol in the barrels suffices for a car to go around the track once.

You have a car but no petrol to start with. However, you may choose the barrel at which you start your journey. Can you choose the place in such a way that you can go full circle, filling up at each barrel when you reach it?

3-4 E 10.19 Suppose in a group of 100 people each person knows at least 50 other people. Prove that it is possible to seat the people around a circular table in such a way that everyone knows his or her two neighbours.

3-4 E 10.20 The general inequality of the geometric, arithmetic and quadratic mean says: Let $n \in \mathbb{N}$ and $a_1, \ldots, a_n \geq 0$. Then

$$\sqrt[n]{a_1 \cdot \cdots \cdot a_n} \leq \frac{a_1 + \cdots + a_n}{n} \leq \sqrt{\frac{a_1^2 + \cdots + a_n^2}{n}},$$

and each inequality becomes an equality if and only if $a_1 = \cdots = a_n$. Prove this.

E 10.21 Among all trapezoids inscribed in a circle, which one has the largest area? 2

E 10.22 Let $n \in \mathbb{N}$. We consider numbers $0 = x_0 < x_1 < \cdots < x_n = 1$. For which values of the x_i does the sum $(x_1 - x_0)^2 + (x_2 - x_1)^2 + \cdots + (x_n - x_{n-1})^2$ take on its smallest possible value? 3

E 10.23 Prove that the equilateral triangle has the largest area among all triangles of a given perimeter. 3-4

E 10.24 Suppose you are given a finite number of points in the plane, not all on a line. Is there always a line which contains precisely two of the points? 4

E 10.25 Let $n \in \mathbb{N}$, $n \geq 2$. Show that on every convex billiard table without corners you can find n boundary points with the following property: a billiard ball[16] shot from one of the points to the next will hit the boundary at precisely the n points in the given order, come back to the initial point and then keep repeating the same track. See Figure 10.11 for an example of an eternal triangular track on an elliptical billiard table. 4

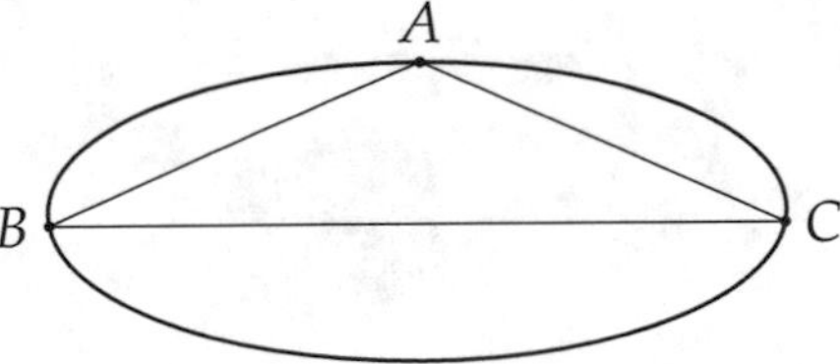

Figure 10.11 An eternal billiard ball track on an elliptical table, $n = 3$ (B and C lie a little below the middle)

[16]We assume that the billiard ball is a point. The claim remains true for 'real' billiard balls, if their radius is smaller than the smallest radius of curvature of the boundary. Also we assume there is no friction.

11 The invariance principle

"When things change, pay attention to what remains constant." While this may sound like ideology, it is really one of the fundamental ideas pervading all of mathematics, and a powerful problem-solving strategy.

Many complex processes are composed of simple steps: from a small number of legal moves the most complex chess games can arise. If you go for a walk in a city then each step is simple, but after a while you may get lost. A physical system changes only slightly in a moment, but after some time it may have changed its state completely, and almost unpredictably.

The invariance principle helps you gain information about such processes in spite of their complexity. The problems in this chapter will show you how and where to use it for solving mathematical problems. Along the way you will find out surprising facts about the 15-puzzle and peg solitaire, and learn about permutations and their signature, fundamental concepts of mathematics.

11.1 The invariance principle, first examples

We start with a few simple examples to show what the invariance principle[1] is and how to use it.

? Problem 11.1

Two joggers A and B run on a straight track. B starts one metre ahead of A. They run at the same speed.

Show that A can never catch up with B.

! Solution

Since A and B have the same speed their distance is constant. Since it is positive at the beginning it will always be positive. !

[1] invariant = not changing

D. Grieser, *Exploring Mathematics*, Springer Undergraduate Mathematics Series, https://doi.org/10.1007/978-3-319-90321-7_11

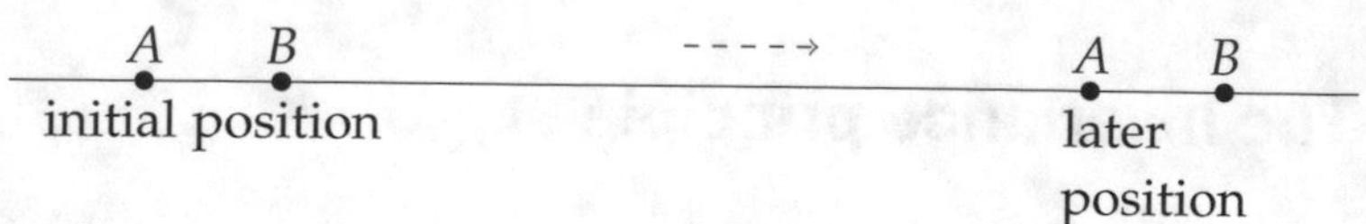

Problem 11.2

You have a board composed of 6×6 *squares. A coin lies on the lower left corner square. In one move the coin can be moved from its present square to one of the squares diagonally adjacent to it.*

Show that the coin can never get into the lower right corner square, no matter how long you play.

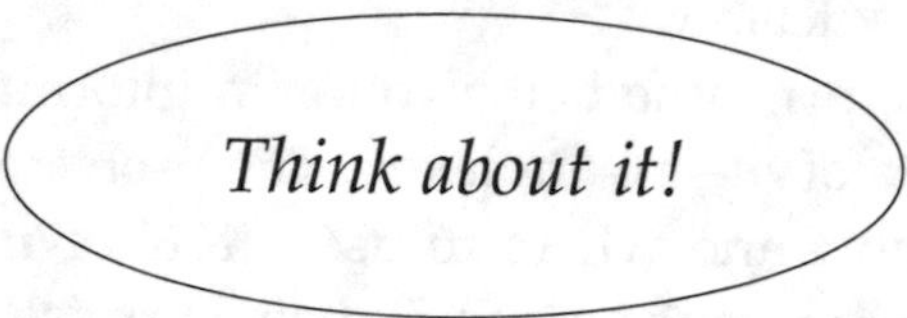

Solution

Colour the squares black and white as on a chessboard. The lower left square is black, say. In each move the coin will remain on the same colour, so it will always remain on black squares. Since the lower right square is white the coin can never get there. The coin is like a (lame) bishop in chess. !

What do the two problems have in common? Something remains constant (in each moment or in each move), therefore it must remain constant over an arbitrary time span or after arbitrarily many moves. This is the invariance principle. In short:

Invariance principle

When things change look out for what remains constant!

The examples show that the invariance principle helps with **proofs of impossibility**: B cannot catch up with A, the coin can never reach the lower right square.

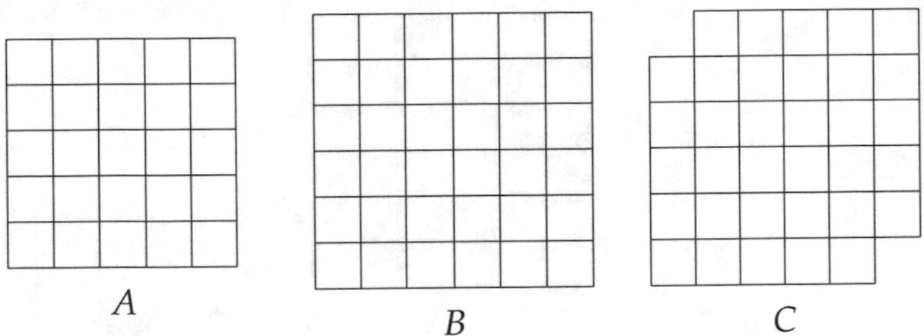

Figure 11.1 Boards for domino tilings

? Problem 11.3

Is it possible to tile a 5×5 *board with dominoes? How about a* 6×6 *board?*

The boards are shown in Figure 11.1 A and B. A domino is a 2×1 rectangle, and **to tile** means to cover without gaps and without overlaps. Each domino covers two adjacent 1×1 squares.

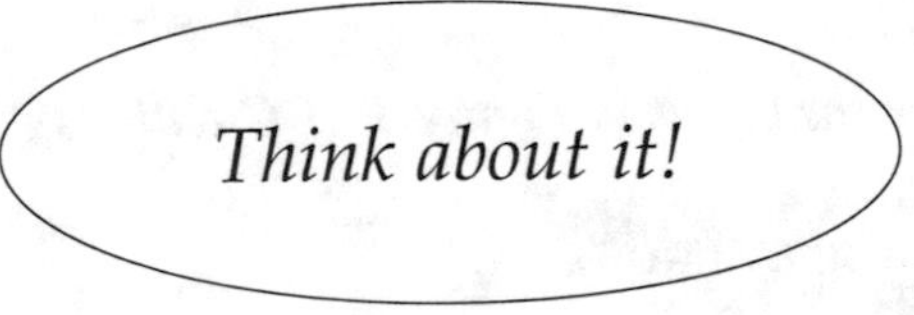

! Solution

A domino covers two squares. Therefore any board tiled with dominoes must have an even number of squares. The 5×5 board has 25 squares, so it cannot be tiled.

The easiest way to show that a 6×6 board can be tiled with dominoes is to give an example of a tiling, see Figure 11.2.

> **Be careful:** For proving that the 6×6 board can be tiled it is *not* enough to argue that it has an even number of squares. For example, the board ⊞ has an even number of squares, but it obviously cannot be tiled with dominoes.

The proof of impossibility for the 5×5 board can be viewed as an instance of the invariance principle: think of putting dominoes on the

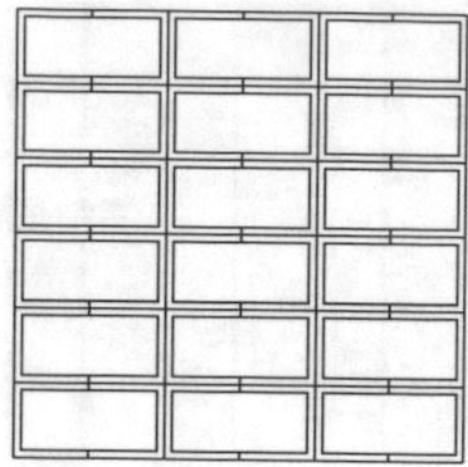

Figure 11.2 Example of a domino tiling

board one by one. The number of covered squares increases by 2 in each step, so its parity stays the same. Since this number starts out at 0 it can never become 25.

Recall that the **parity** of an integer is its property of being even or odd. Parity is often a good candidate for an invariant, but sometimes it does not help, as the next example shows.

? Problem 11.4

Is it possible to tile the board in Figure 11.1 C with dominoes?

Investigation and solution

▷ The board has $36 - 2 = 34$ squares, an even number. So it might be possible to tile it, but we cannot be sure. In any case we cannot use the parity argument to prove that it is impossible. Try to find a tiling!

▷ You didn't manage to tile the board? Then we formulate the **conjecture:** It is impossible.

▷ How could we prove the conjecture?

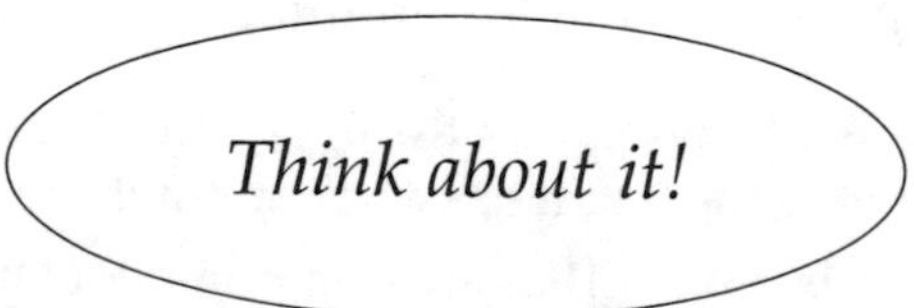

We could try out all possibilities – but there is no hope of doing that, there are just too many, and there is no obvious way to check them systematically.

▷ Therefore we should **look for a new invariant**: some quantity that remains the same when we add a domino to the board. In addition, this quantity should have different values for the empty board and the (hypothetically) completely tiled board, otherwise it won't be useful for our proof of impossibility.

The parity of the number of covered squares is an invariant, but it does not satisfy the additional condition since 0 and 34 are both even.

▷ Idea: We use the chessboard colouring from above. A domino will always cover a white square and a black square, *no matter where you place it.* Therefore any board tiled with dominoes must have the same number of white and black squares.

▷ The two corner squares which we removed have the same colour (say white). Therefore our board has more black squares than white squares, so it cannot be tiled with dominoes. !

Review of Problem 11.4

The essential idea was to introduce the chessboard pattern. This was not present in the original problem. Often it is useful **to introduce additional structure.** This has already helped us to solve Problem 11.2.

Summary on tilings: In order to decide whether the boards A, B, C in Figure 11.1 can be tiled with dominoes we argued as follows:

- Using the invariant 'parity' we proved that board A cannot be tiled.
- For boards B and C the parity argument is useless. It does not help us decide whether a tiling is possible.
- Board B can be tiled. We proved this by giving an example of a tiling.
- Board C has no tiling. We proved this using another invariant: the difference of the numbers of covered white and black squares.

11.2 How to use the invariance principle

Now we can describe more precisly how to use the invariance principle.

Problem: Suppose we have a problem which involves a **process** which advances by **steps** through various **states.** See Table 11.1. Our goal is to understand which states can be reached from a given initial state. It is known which steps are possible from any state. Usually a single step is very simple, but after several steps things get complicated.[2]

An **invariant** of the process is a rule assigning to each possible state a number (or a symbol, a colour, a word etc.), which remains constant during each step.

	jogger	coin on square	domino tilings
process	running	moving the coin	tiling the board
step	–	move the coin to adjacent square	add a domino to the board
state	positions of the joggers	position of the coin	tiling of part of the board
invariant	distance of the joggers	colour of the square that the coin is on	parity of the number of covered squares; difference white/black covered squares

Table 11.1 Problems 11.1–11.4

Preparation: We familiarise ourselves with the problem (using examples etc.) and arrive at a *conjecture* about which states can be reached and which cannot.

Proof of existence for reachable states: The simplest way to show that a state can be reached is to establish a concrete sequence of steps leading to this state.

[2]The process in the jogger problem 11.1 is continuous: it does not have single steps. However, the question and the solution are analogous. See also Section 11.4.

Proof of nonexistence for non-reachable states In order to show that a state a cannot be reached we look for an invariant of the process which takes different values at the initial state and the state a.

As usual it may happen that we need to correct our initial conjecture, or that an invariant which we are trying to use does not help. And as usual the name of the game is: **try again, don't give up!**

As for the extremal principle, you will need to use your **creativity** to find a suitable invariant. There are no limits: **if it works then it is permitted.**

We can formalise the problem and the invariance principle as follows:[3]

We are given a set M, the set of states, and a subset $S \subset M \times M$, the set of steps.
(Interpretation: $(a, b) \in S$ means that one can get from state a to state b in one step.)

We say that a state $b \in M$ is **reachable** from a state $a \in M$ if there is a run of the process starting with a and ending with b. Here a **run** is a sequence $a, x_1, x_2, \ldots, x_n, b$ of elements of M for which $(a, x_1), (x_1, x_2), \ldots, (x_n, b) \in S$, for some $n \in \mathbb{N}$.

An **invariant** is a map $I : M \to T$ into an arbitrary set T which satisfies: for all steps $(a, b) \in S$ we have $I(a) = I(b)$.

Invariance principle: If I is an invariant and $a, b \in M$ satisfy $I(a) \neq I(b)$ then b is not reachable from a.

Example

In Problem 11.2 M is the set of squares of the board, S is the set of pairs (a, b) of squares a, b which are diagonally adjacent to each other, $T = \{\text{black}, \text{white}\}$ and $I(a)$ is the color of square a in the chessboard colouring, for each $a \in M$.

[3] This formalisation is not needed for understanding the rest of the chapter. It serves to make the concepts precise, and illustrates the way many mathematical texts are written. However, it will not usually help you to get good ideas for finding invariants. For that it is more important to have an intuitive feeling for the invariance principle, and you get this by doing examples, solving problems. Conversely you may conclude that you should always look for the idea behind the formal definitions that you find in many textbooks on mathematics.

11.3 Further examples: number games, permutations, puzzles and peg solitaire

Often we describe processes as games. Then the states are the possible positions of the game and the steps are the moves that are possible in a given position.

Problem 11.5

Let n be an odd natural number. We play the following game: We write down the numbers $1, 2, \ldots, 2n$. A move consists of choosing two numbers and replacing them by their difference. We play until only one number is left.

Can the remaining number be zero?

Investigation

▷ Let us consider an **example** with $n = 3$. We mark the two numbers chosen for the next move by a dot:

$$\dot{1}\ 2\ 3\ 4\ 5\ \dot{6}$$
$$2\ 3\ 4\ \dot{5}\ \dot{5}$$
$$\dot{2}\ \dot{3}\ 4\ 0$$
$$\dot{1}\ \dot{4}\ 0$$
$$\dot{3}\ \dot{0}$$
$$\dot{3}$$

In this case the number 3 remains. We could have proceeded differently; for example we could have chosen 1 and 2 in the first step. Play a few different runs of the game. You will notice that you never get zero at the end. So we formulate:
Conjecture: For $n = 3$ the remaining number cannot be zero.

▷ How could we prove that? Certainly we should not try out all possibilities. There are too many (and in any case we want to understand the problem for *any* odd n).

▷ The form of this problem matches exactly the scheme described above: we have a process (the game) consisting of single steps (the moves), and we want to show that a certain final state is impossible. So we should **look for an invariant.**

▷ What are the states? They are finite sequences of numbers: the lines in the example above. So we look for a rule assigning to each such sequence a number (or colour or ...), which remains constant during each move, *no matter what move we make.*

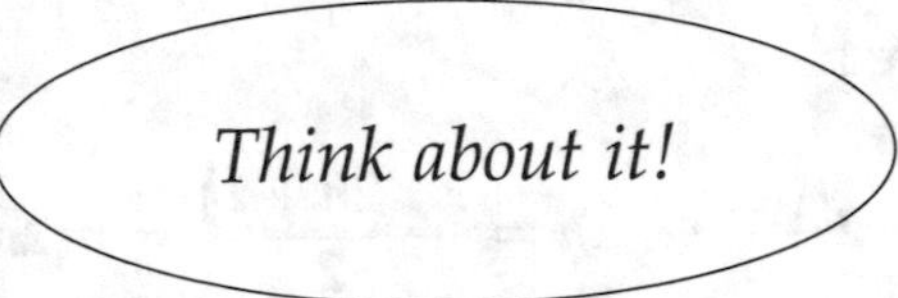

▷ One of the simplest things you can do with a finite sequence of numbers is to add all the numbers. Let's try this in the example:

sequence	sum
$\dot{1}\ 2\ 3\ 4\ 5\ \dot{6}$	21
$2\ 3\ 4\ \dot{5}\ \dot{5}$	19
$\dot{2}\ \dot{3}\ 4\ 0$	9
$\dot{1}\ \dot{4}\ 0$	5
$\dot{3}\ \dot{0}$	3
$\dot{3}$	3

Do you notice something? The sums are always odd. We get an idea:
Conjecture: The parity of the sum of the numbers is an invariant.
If we could prove this then we would be done for $n = 3$ since the sum starts out as 21, so it can never become zero. Then for the general case we only need to check whether the sum $1 + 2 + \cdots + 2n$ is always odd if n is odd. Try it yourself!

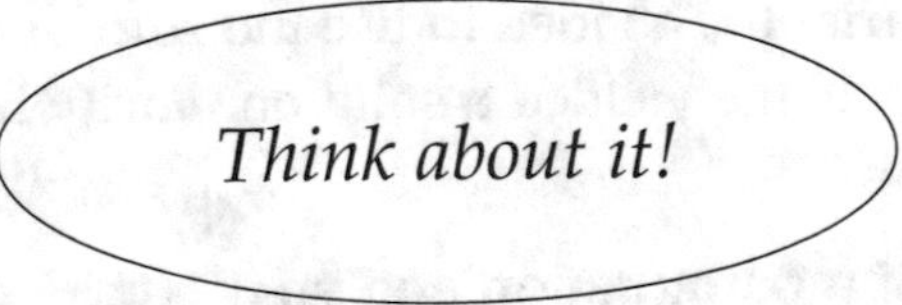

Solution of Problem 11.5

The number remaining at the end cannot be zero. In order to show this we prove two propositions:

1. The sum $1+2+\cdots+2n$ is odd.

2. The parity of the sum of the remaining numbers stays the same during each move.

1. and 2. together imply that the single number remaining at the end must be odd, so it cannot be zero.

Proof of 1.: We have

$$1+2+\cdots+2n = \frac{2n(2n+1)}{2} = n(2n+1).$$

By assumption n is odd. Since $2n+1$ is also odd, it follows that $n(2n+1)$ is odd.

Proof of 2.: Suppose at some stage in the game the numbers $a_1,\ldots,a_k,x,y$ remain, where x,y are the numbers chosen for the next move. Here $k \in \mathbb{N}_0$ could be zero. Without loss of generality we may assume $x > y$. Then after the next move the numbers $a_1,\ldots,a_k,x-y$ remain. So we need to show

$$a_1+\cdots+a_k+x+y \equiv a_1+\cdots+a_k+x-y \mod 2$$

(for the congruence notation see Chapter 8; this just means the two numbers have the same parity). The congruence is true since the difference of the left- and right-hand sides is $2y$, hence even. !

Review of Problem 11.5

The setup (moves etc.) and our conjecture of impossibility suggested that we use the invariance principle. But it was not obvious what to use as an invariant. The idea to use the sum of the numbers at a given stage of the game yielded a solution. Parity helped, as is often the case.

The signature of a permutation and an 8-puzzle problem

A **permutation** is an ordering of the numbers $1,2,\ldots,n$. For $n=3$ there are 6 permutations: $123, 132, 213, 231, 312, 321$. We have already encountered permutations in Chapter 5, where we determined that there are $n!$ of them.

The following problem will guide you to the notion of the signature of a permutation, one of the fundamental invariants of mathematics.

? Problem 11.6

Write down the numbers 1, 2, ..., n *in any order. A move swaps any two adjacent numbers. Is it possible to arrive at the initial order after an odd number of moves?*

Investigation

▷ Again we are talking about a process: the states are the permutations, the moves are the neighbour swaps.

▷ Try out a few **examples**. For example, starting with 213 we could swap 1 and 3 and arrive at 231; then we could continue as 321, 312, 132, 123, 213, which is the order we started with. We have made six moves. You will notice that whenever you arrive back at the initial order you will have made an even number of moves. We **conjecture** that this is always the case. How could we prove that in general?

▷ It is not clear how to use the invariance principle here. But if we could assign a number to each permutation in such a way that this number *changes its parity with each move* then our conjecture would follow immediately. This is our **plan.**

▷ What happens when we swap two adjacent numbers? Two numbers that were in natural order (i.e. the smaller number first) will now be out of order, and conversely.

▷ Idea: this suggests that we consider the number of **inversions** of a permutation. An inversion is a pair of places $i < j$ where the number at place i is bigger than the number at place j.
Example: How many inversions does the permutation 3412 have? We consider all pairs of places systematically: 34 (at places 1 and 2) is not inverted, 31 inverted, 32 inverted, 41 inverted, 42 inverted, 12 not inverted. So there are 4 inversions.

▷ How does the number of inversions change when we swap two adjacent numbers? Clearly, if these numbers were not inverted

before the swap then they will be afterwards, and if they were inverted then they won't be afterwards. The position of these two numbers relative to all other numbers remains unchanged, as do the positions of any pair of the other numbers.
So the number of inversions changes by one when we swap two adjacent numbers. In particular the number of inversions will change its parity, which is what we wanted. Here is an example. The signature is the parity of the number of inversions.

permutation	number of inversions	signature
2 1 3	1	o
2 3 1	2	e
3 2 1	3	o
3 1 2	2	e
1 3 2	1	o
1 2 3	0	e
2 1 3	1	o

Figure 11.3 Example for Problem 11.6; o = odd, e = even

▷ At the beginning and the end we have the same permutation, so the number of inversions is the same. Therefore the parity of the number of inversions must have changed an even number of times; that is, we must have made an even number of moves.

We summarise the definitions and write up the solution neatly.

Definition An ordering of the numbers $1, 2, \ldots, n$ is called a **permutation** of $1, \ldots, n$. We write it as[4] $(a_1, a_2, \ldots, a_n)$. An **inversion** is a pair of numbers $i, j \in \{1, \ldots, n\}$ satisfying

$$i < j \quad \text{and} \quad a_i > a_j.$$

The **signature** of the permutation $(a_1, \ldots, a_n)$ is the parity of the number of its inversions. We call the permutation **even** or **odd** if its signature is even or odd, respectively.

Often we denote permutations by small Greek letters, for example π (pi) or σ (sigma).

Solution of Problem 11.6

We first prove a lemma.

Lemma Swapping two adjacent numbers in a permutation changes its signature.

Proof.
Let $\pi = (a_1, \ldots, a_n)$ be a permutation. Let $1 \le k < n$ and $\sigma = (a_1, \ldots, a_{k+1}, a_k, \ldots, a_n)$ be the permutation obtained after swapping a_k and a_{k+1}. We show that the number of inversions of σ is one bigger or one smaller than the number of inversions of π. This implies the lemma. For this we consider the possible inversions $i < j$ of π and σ:

- If neither i nor j lie in $\{k, k+1\}$ then σ has the same numbers at positions i, j as π, so i, j is an inversion of σ if and only if it is an inversion of π.
- Any inversion of π with $i = k$ and $j > k+1$ will lead to an inversion of σ with $i = k+1$ and the same $j > k+1$ since a_k moves to position $k+1$ in σ. Similarly, any inversion of π with $i = k+1$ and $j > k+1$ leads to an inversion of σ with $i = k$ and $j > k+1$.

  Therefore σ and π have the same number of inversions with $i \in \{k, k+1\}$ and $j > k+1$.
- Similarly, σ and π have the same number of inversions with $i < k$ and $j \in \{k, k+1\}$.
- It remains to consider the case $i = k$, $j = k+1$. If π has an inversion at the position pair $k, k+1$, i.e. if $a_k > a_{k+1}$, then σ does not have an inversion at this position pair. If π does not have an inversion there then σ does.

[4]You can also view a permutation $(a_1, \ldots, a_n)$ as a bijective map $\{1, \ldots, n\} \to \{1, \ldots, n\}$, $n \mapsto a_n$. For example, the permutation $(2, 1, 3)$ corresponds to the map $1 \mapsto 2, 2 \mapsto 1, 3 \mapsto 3$.

All in all, we see that the number of inversions of σ differs from the number of inversions of π by 1. q. e. d.

We can now solve Problem 11.6: Since the signature changes at each move, and is the same at the beginning and the end, the number of moves must be even. !

Remark

The signature is not an invariant but a **semi-invariant** for this problem: while it does not remain constant, it changes in a simple way with each move, and the change remains predictable even after many moves. This was sufficient for solving the problem.

We note two other important properties of the signature. You should prove them as Exercise E 11.6.

Theorem **(Properties of the signature)**

a) When you move a number in a permutation to the left or right, skipping k other numbers, then the signature changes if k is odd and remains the same if k is even.

b) When you swap *any* two numbers in a permutation then the signature changes.

Example for a): 31254 is odd, and if you move 1 to the right, skipping 2 and 5, you obtain 32514, which is also odd.
Part b) generalizes the previous lemma on swapping adjacent numbers.

Here is a beautiful application of the signature:

? Problem 11.7

Consider a sliding puzzle in the shape of a 3×3 *square, filled with 8 square tiles numbered 1 to 8 as in Figure 11.4, so one tile is missing. A move slides an adjacent tile into the empty space.*

Somebody takes out the tiles [7] *and* [8] *and puts them back in the opposite order. Can you get back from this arrangement to the original arrangement by sliding moves?*

This puzzle is called the **8-puzzle.** Try it!

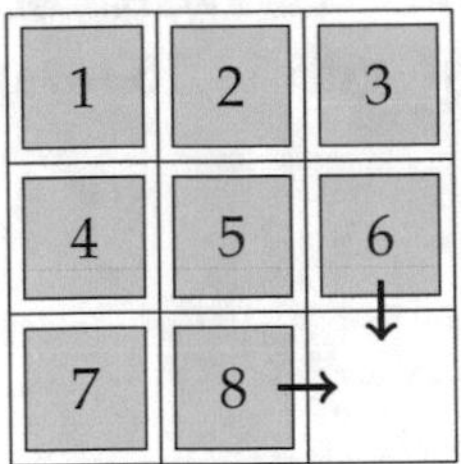

Figure 11.4 Left: original position of the 8-puzzle, with two possible moves; right: tiles 7 and 8 are interchanged

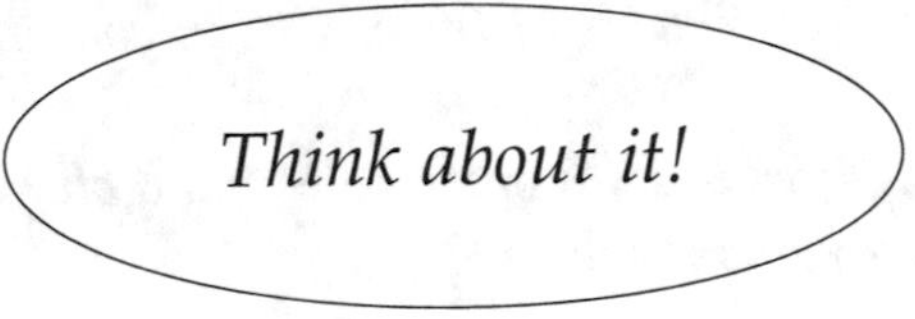

! Solution

The states are all possible arrangements of the tiles. By reading the numbers row by row and ignoring the empty space we can associate a permutation of the numbers $1, 2, \ldots, 8$ with each arrangement. The original arrangement corresponds to the permutation 12345678; the arrangement on the right in Figure 11.4 corresponds to the permutation 12345687.

When we move a tile horizontally then the corresponding permutation does not change. Let us see what happens when we move a tile vertically: Let m be the number on the tile. If we slide m down then in the corresponding permutation it moves right, skipping two numbers. If we slide m up then it moves left, skipping two numbers. By the theorem above the signature does not change, whichever way we move m.

Therefore the signature of the associated permutation is an invariant of the game. Since the permutations 12345678 and 12345687 have different signatures it is impossible to get from one arrangement to the other, no matter how many moves we make. !

The analogous problem for a 4×4 square (the so-called 15-puzzle, see Exercise E 11.9) has an interesting history. It is reported that in the year 1880 the puzzle created a craze first in the US and then in Europe, after someone had promised a 1000 Dollar award for finding a sequence of moves from the arrangement with 14, 15 swapped to the original arrangement.

Three-coloured balls and peg solitaire

? Problem 11.8

You have a box containing 4 red balls, 5 green balls and 6 blue balls. In each move you take out any two balls of different colours and replace them by a ball of the third colour.[5] *You play until you cannot move any more.*

Suppose a single ball remains at the end. Can you predict which colour it is?

If you start with 4 red balls, 6 green balls and 8 blue balls, is it possible that a single balls remains at the end?

Try it. Observe carefully the numbers of red, green and blue balls in the course of the game. Do you notice anything? Can you prove it? What can you conclude from it?

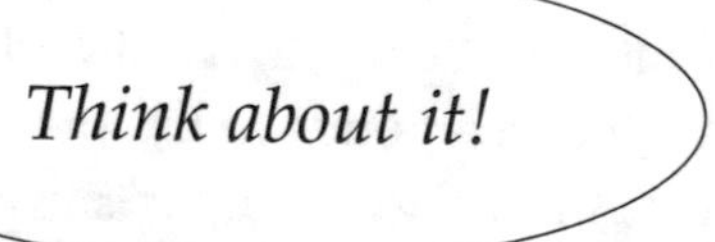

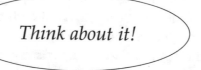

! Investigation and solution

▷ Denote the numbers of red, green and blue balls by r, g, b, respectively. Let us observe what happens to r, g during a move: if you take out red - green then they change to $r-1, g-1$. If you take out red - blue then they change to $r-1, g+1$. And if you take out green - blue then they change to $r+1, g-1$. Can you name something that does not change in each case?

[5] It is assumed that you have an extra supply of balls of each colour outside the box.

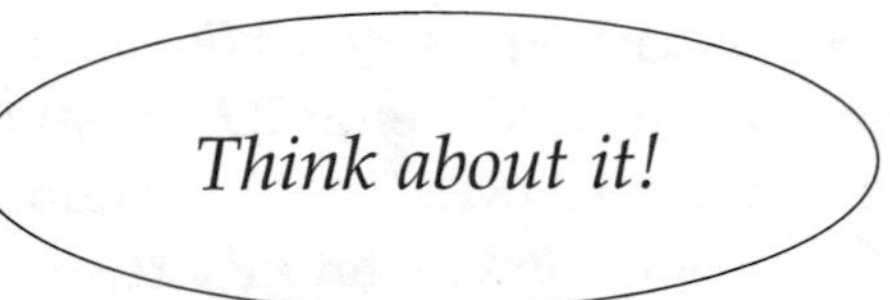

▷ If r, g have the same parity before the move then they have the same parity after the move. And if they have different parity before the move they have different parity after the move. Put differently, the parity of their difference does not change, *no matter which move we make*.

▷ By symmetry the same is true for r, b and for b, g: what remains constant is whether they have the same or opposite parity.

▷ Suppose a single red ball remains at the end. Then r, g, b are $1, 0, 0$ at the end, so r has different parity from g and b. Therefore, r must have had different parity from g and b at the beginning.

Similarly, if a green or blue ball remains at the end then g or b respectively must have had a different parity from the other two numbers at the beginning.

▷ Since the starting values of r, g, b are $4, 5, 6$ and only 5 has a different parity from the other two numbers, it follows that the remaining ball is green.

▷ In the case $4, 6, 8$ all parities are equal at the start, so they must always be equal. Therefore it is not possible that a single ball remains at the end.

Remarks

- We did not show that in the case $4, 5, 6$ it is actually possible to choose the moves so that a single ball remains at the end (which we proved would have to be green). This was not part of the question. On the other hand, you will easily find a way to achieve this. However, this is not completely trivial: if you start badly then you can manoeuvre yourself into a dead end. For example if you first take out green - blue you get $5, 4, 5$; if you now take red - blue five times then you get $0, 9, 0$, and the game is over.

- Our solution yields a *necessary condition* for the possibility of leaving one ball only at the end: the initial numbers must be not all odd, and not all even. Is this condition also sufficient? (No, see the previous remark and Exercise E 11.19).

- This pleasing problem has more to offer: it can be used to give surprising insights into the famous game of **peg solitaire.** For example: if you manage to leave a single peg anywhere, then you could easily have left a single peg in the centre instead. See Exercise E 11.23.

11.4 Going further: knots, conserved quantities, and why you should care about proofs of impossibility

Invariants of continuous processes

The process in the jogger problem 11.1 is not composed of discrete steps. Its state changes continuously. This is best modelled using notions of mathematical analysis. In this context the invariance principle is a theorem of analysis:

> If the derivative of a function is zero everywhere then the function is constant.

More precisely: Suppose the function $f : [a,b] \to \mathbb{R}$ is differentiable and $f'(x) = 0$ for all $x \in [a,b]$. Then f is constant, in particular $f(a) = f(b)$. Loosely speaking: if f doesn't change "in the small" anywhere – that is, its graph has only horizontal tangents – then it does not change "in the large", i.e. from a to b.

In the jogger problem we take $f(x) = B(x) - A(x)$ to be the distance between the joggers, where x is time and $A(x)$, $B(x)$ are the locations of the joggers, measured along the number line, at time x. The speeds of the joggers at time x are the derivatives $A'(x)$, $B'(x)$, and since these are assumed to be the same we have $f'(x) = B'(x) - A'(x) = 0$ for all x, so f is constant.

This continuous invariance principle is the basis for the proof that solutions of many differential equations are unique.

Examples of the invariance principle in various areas of mathematics

The idea of invariants pervades all areas of mathematics. Here are some examples.

As in the case of the extremal principle in Section 10.4, you may encounter here some words of higher mathematics that you don't know yet. But keep reading anyway; you will meet them again when you learn more maths, and then remember that they fit into this grand scheme.

- In Section 4.5 you learned about the EULER **characteristic**. This is an invariant of topological spaces – these are for example surfaces in space, like a sphere or a torus. The EULER characteristic is invariant are deformations of these spaces. These are continuous processes.
- For a plane curve which is closed (i.e. starting point and endpoint are the same) and which is allowed to intersect itself, and for a point P not lying on the curve, one can consider the winding number: the number of times the curve "winds around" P. The winding number is an invariant of such curves under deformations, assuming that during the deformation the curve always avoids P. It is not easy to turn this intuitive description of the winding number into an exact definition. This can be done using tools from topology or from complex function theory.

  The winding number has many applications. For example it can be used for a very elegant proof of the fundamental theorem of algebra (any polynomial has a zero in the complex numbers).[6]
- A very appealing and important application of invariants is the investigation of **knots.** The rope on the left in Figure 11.5 can easily be unknotted, while the second one – the so-called trefoil knot – cannot; at least that's what our experience tells us. But can we be sure, have we tried all possibilities?[7] And how about the third and fourth rope? How can we decide generally whether a

[6]See the Wikipedia entry for 'Fundamental theorem of algebra'.

[7]Remember Problem 4.5!

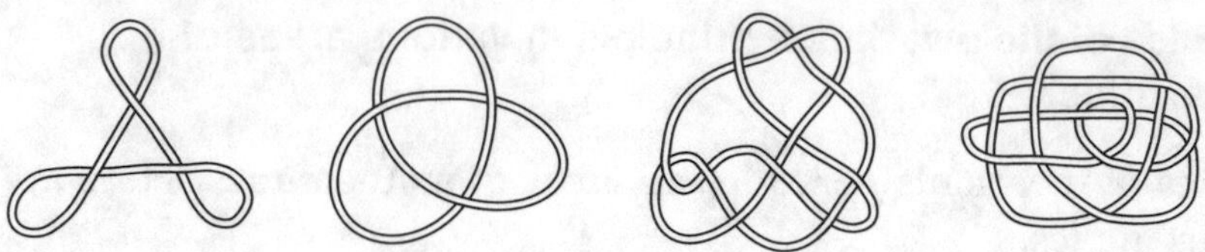

Figure 11.5 Four 'knots'

given closed rope can be unknotted or not?[8] Can one *prove* that the three knots on the right can not be unknotted?

These questions are investigated in **knot theory,** a fascinating modern mathematical theory which has applications for example in the understanding of DNA, which encodes our genetic information. In knot theory one constructs invariants which are easy to compute and which can be used to prove, for example, that the second and third knot in Figure 11.5 cannot be unknotted.

However, up to this day no one has succeeded in proving that these invariants are sufficient for deciding whether any given knot can be unknotted. We don't know whether they are.

An even more ambitious goal would be to find a **complete set of invariants,** which would enable us to decide for any two knots whether they can be deformed into each other.[9]

An extension cable is useful for playfully investigating knots (knot it, then plug the ends together).

By the way, the knot on the right in Figure 11.5 can be unknotted.

❑ Processes and their long-term behaviour are the subject of the theory of **dynamical systems.** Important tools in this theory are invariant manifolds and LYAPUNOV functions (these are semi-invariants: they are monotone increasing or decreasing during a run of the process).

[8] We make this more precise: A **knot** is a closed curve in space which does not intersect itself. Two knots are called **equivalent** if one can be deformed into the other, where also during the deformation no self-intersections are allowed. A knot can be unknotted (alternatively, it is called **trivial**) if it can be deformed to the **unknot**, i.e. a circle.

[9] The book (Adams, Colin C., 2004) gives an informal introduction to knot theory, with applications in physics, chemistry and biology; on the web page `http://katlas.org/wiki/Main_Page` you find pictures of many knots.

- The invariance principle is also of fundamental importance in **physics.** For example, the law of conservation of energy says: the energy of any closed system remains constant, no matter how the system changes. Such physical invariants are usually called **conserved quantities.** Other conserved quantities are momentum and angular momentum. Using these invariants skilfully will enable you to find simple solutions for some apparently complex problems.[10] Albert EINSTEIN discovered (special) relativity by searching for a theory in which the laws of nature are described in a LORENTZ invariant way and in which the speed of light is constant.

Invariance is often understood in a broader sense, not just referring to constancy in a process. Here are some examples which you meet quite early in university mathematics.

- Invariance under maps: Modern mathematics is usually formulated in the language of sets and maps, since this gives it a unified basis. Sets may carry additional structure. For example, there are vector spaces, groups or topological spaces. Then the invertible maps which preserve the extra structure, so-called isomorphisms, are of special importance. In the examples they are linear isomorphisms, group isomorphisms and homeomorphisms, respectively. An invariant of a structure is a quantity (or some other mathematical structure, for example a group) which can be associated with any instance of the structure, and which is the same for two instances if there is an isomorphism between them.

  For example the dimension is an invariant of vector spaces, and the EULER characteristic is an invariant of topological spaces.[11]

- Invariance as independence of a result (of a calculation) from the details of its derivation. For example, we can find the dimension

[10]Example: Suppose an object falls down, starting at height h. What is its velocity v when it hits the ground? The easiest way to compute v is by conservation of energy: potential energy lost = kinetic energy gained, so $mgh = \frac{1}{2}mv^2$, hence $v = \sqrt{2gh}$, where g is the gravitational acceleration of the earth. This method also works if the object slides down a slope (without friction) – no matter how steep the slope is, and even if it has variable slope, the result is always the same!

[11]This implies in particular the invariance of the EULER characteristic under deformations that we talked about in Section 4.5.

of a vector space by choosing a basis and counting its vectors – any other basis would yield the same result. The EULER characteristic of a surface can be computed by drawing a graph on the surface and counting its vertices, edges and faces (see Section 4.5 for details) – the result is independent of the choice of graph. Determinant and trace of a linear map from a vector space to itself are computed by taking the determinant or trace of a matrix representing the map – the result is independent of the choice of basis used to compute that matrix. In differential geometry many quantities, for example curvature, can be defined and computed using local coordinates. Such quantities will only be geometrically meaningful if their calculation in a different coordinate system yields the same result.

We could go on and on. Keep your eyes open!

Summarising, we can say that invariance is one of the fundamental ordering principles of mathematics – in addition to its use as a problem solving strategy which you learned about in this chapter.

Why should you care about proofs of impossibility?

If we prove that something is impossible then we can stop spending energy on trying to do it. Such knowledge improves our understanding of the subject, and we may go on to modify the question or take a different point of view.

There is also a very practical side to proofs of impossibility. Cryptography provides a good example: Many a convenience of everyday life relies on the possibility of encrypting information: using cash cards, shopping on the internet etc. A common algorithm for encryption is the RSA algorithm. It is safe as long as no one has found a fast way to decompose very large numbers (several hundred decimal digits) into their prime factors.[12]

But up to now nobody has succeeded in *proving* that no fast way to do this exists. Without such a proof we have to live with the residual uncertainty that someone may find a fast way – or has found it already, but just hasn't told us yet. A proof of impossibility would

[12]'Fast' is not a mathematical but rather a practical notion. You may think of 'in less than a thousand years on state-of-the-art computers'.

be very welcome for us all here, besides being a big breakthrough in research.

11.5 Toolbox

The invariance principle is a fundamental tool for proving that something is **impossible** or **non-existent:** If you can reformulate the problem in terms of a process then you should **look for invariants.** To find an invariant that yields a desired proof of impossibility or non-existence may be difficult, and may require **creativity** and many attempts. You also learned about the **signature of a permutation** which can be useful as an invariant or semi-invariant. In Problem 11.4 you learned about another important strategy: **introducing additional structure** (like a chessboard pattern), that is, a structure that is not present in the problem but helps with its solution.

Exercises

E 11.1 We shred a sheet of paper. In each step we take one piece and tear it into 3 or 5 parts. Is it possible to obtain precisely 100 pieces? 1

E 11.2 To show that there is no tiling of a 5×5 square with dominoes we used a parity argument. Can you show this using a chessboard colouring instead? 1

E 11.3 We colour the board in Figure 11.1 C as follows: the lower left 3×3 square and the upper right 3×3 square are black, the rest is white. Can you use this colouring instead of the chessboard colouring to prove that the board cannot be tiled with dominoes? 1

E 11.4 Let $n \in \mathbb{N}$ be odd. Write down the numbers $1, 2, \ldots, n$. In each step delete any three numbers and replace them by their sum. Show that after some time only one number is left, and determine this number. 1-2

E 11.5 Consider Problem 11.6 about permutations again. Since only adjacent numbers are swapped in each step you might find it more natural to count inversions of adjacent numbers only, that is, pairs 2

$i, i+1$ having $a_i > a_{i+1}$. Can you solve the problem using this number instead of the total number of inversions?

2 E 11.6 Prove the theorem about the properties of the signature, stated before Problem 11.7.

2 E 11.7 Find a bijection between the set of even permutations and the set of odd permutations of $1, \ldots, n$. How many even permutations are there?

2 E 11.8 A chocolate bar consists of four by six squares. You want to break it into single squares. You are only allowed to break along the lines between the squares, and in each step you can only break up one of the pieces you have made so far. How many steps do you need at least? How many at most?

3 E 11.9 Solve the analogue of Problem 11.7 for the 4×4 square where initially the tiles 14 and 15 are swapped.

2-3 E 11.10 From a 12×12 board you remove one little square from each of three corners. Is it possible to tile the remaining board using tiles of the form □□□?

3 E 11.11 A bathroom floor is tiled with 1×4 and 2×2 tiles. One of the 1×4 tiles breaks, but we only have a 2×2 tile to replace it. Is it possible to rearrange the tiles so that they still completely tile the bathroom floor?

2 E 11.12 Consider the boards in Figure 11.6.

a) Show that it is impossible to tile the right-hand board (where three squares have been removed) with tiles of the form □□□.

b) Prove that you can tile the left-hand board with tiles of the form □□□.

c) Is it possible to tile the left-hand board with tiles of the form □□□□ ?

2-3 E 11.13 Consider a strip of $2n$ unit squares. The squares are coloured alternately black and white. In each move choose a contiguous group of squares (for example the third square, or squares two to four) and swap all colours in this group. Clearly you can make all squares white using n moves. Is this also possible with fewer moves?

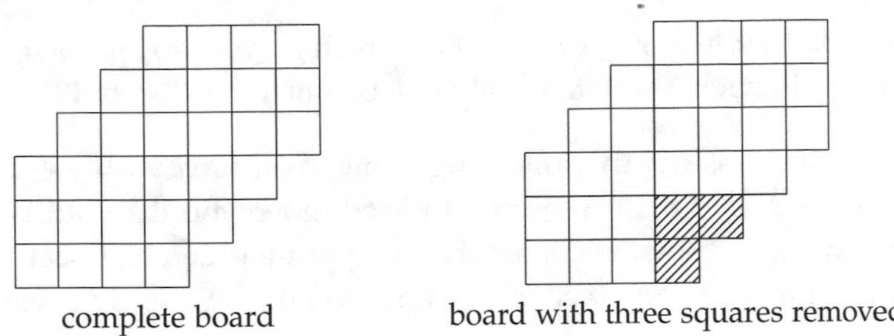

Figure 11.6 Boards for Exercise E 11.12

E 11.14 A heavy square armchair can only be moved by turning it around one of its corners by 90 degrees. Can you move it multiple times so that in the end it is next to its original position, and the back of the chair faces in the same direction as at the beginning? 2-3

E 11.15 The numbers 1,0,1,0,0,0 are written round a circle in this order. In each move you are allowed to add 1 to two neighbours or to subtract 1 from two neighbours. Can you ever make all numbers equal zero? For which initial arrangements of integers round a circle is it possible to do this? 3

E 11.16 Three frogs sit at the points of the plane with coordinates $(0,0)$, $(0,1)$ and $(1,0)$. Now they start leaping. In each leap one frog leaps across another frog so that the leaped-over frog sits in the middle between the starting and landing positions of the leaping frog. Can a frog ever land at position $(1,1)$? (These are super-frogs: they can leap a long way.) 3

E 11.17 Starting with the numbers $1,2,\ldots,4n-1$ replace any two numbers by their difference. Repeat until only one number is left. What can you say about its parity? 2

E 11.18 Let P be a convex n-gon where n is even. Let p be a point in its interior which does not lie on any diagonal. We consider the number of triangles whose vertices are vertices of P and which contain p. Find this number for some examples of P and p. What do you observe? Proof? 2

E 11.19 Consider Problem 11.8 with initially r red, g green and b 2-3 blue balls, where $r, g, b \in \mathbb{N}_0$. For which numbers r, g, b is it possible to play in such a way that only one ball is left at the end?

2-3 E 11.20 Consider the following game. You start with $g \in \mathbb{N}$ green and $r \in \mathbb{N}$ red balls in a box. In each move you take out two balls. If they have the same colour then you put a green ball back into the box, otherwise a red ball. We assume you have a supply of extra balls as needed.

You play until only one ball is left in the box. Can you predict from r and g what colour this ball is?

2 E 11.21 Imagine a museum with one hundred identical square rooms, with a floor plan as in Figure 11.7, where the heavy lines are doors. The entry is at the bottom left, the exit is at the top right.

Is it possible to enter the museum, visit each room exactly once and then leave?

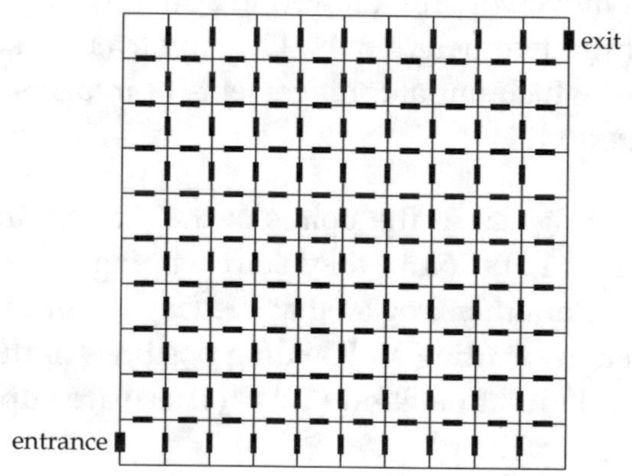

Figure 11.7 The museum in Exercise E 11.21

3 E 11.22 An island is inhabited by a kind of chameleon which can take on the colours red, green and blue. Whenever two chameleons of different colour meet then they will both change their colour to the third, remaining colour. Suppose the numbers of red, green and blue chameleons are $r, g, b \in \mathbb{N}_0$ initially. For which values of r, g, b can it happen that after a while all chameleons have the same colour? For which values is this colour red?

E 11.23 **Peg solitaire** is a game played on the board shown in Figure 11.8. The circles are holes in the board. They are occupied by a peg ◉ or are empty ○. In each move you take a peg and jump horizontally or vertically over an adjacent peg, landing on the empty hole directly behind it. Then you remove the jumped peg. So in the initial position shown on the left in the figure, four moves are possible, and after any of them there will be two empty holes. The game ends when you cannot move any more. 3-4

The aim is to play in such a way that only one peg is left at the end.

a) Show that in a successful game the remaining peg can only be at one of the five positions shown in the middle figure. Analyse the statement "I managed to leave one peg at the end, but unfortunately it was not in the center."

b) Show that no successful game is possible if you start with the initial position shown in the right-hand figure.

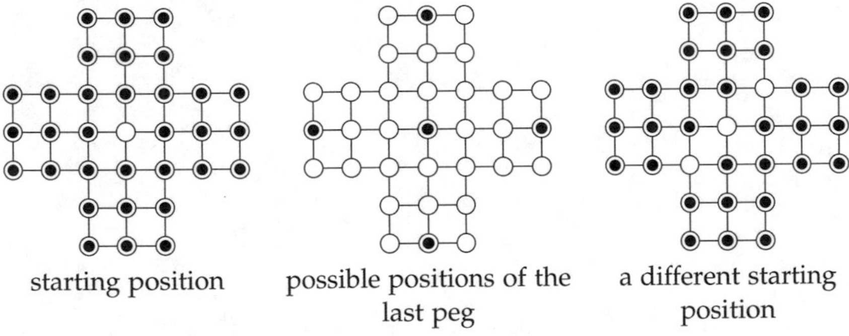

Figure 11.8 Peg solitaire, exercise E 11.23

E 11.24 **Infinite solitaire** is played like solitaire but on an infinite board: there is a hole at each lattice point (n, m), $n, m \in \mathbb{Z}$, of the plane, and initially all holes in the lower half plane are occupied by pegs, that is all holes at points where $m \leq 0$. The aim is to advance as far as possible into the upper half plane. So you want to get a peg to a position (n, m) with big m. What is the largest possible m? 4

E 11.25 In this problem it is assumed that you know **Rubik's cube.** From the outside you can see 26 little coloured cubes called cubelets. When you play with the cube you will notice that from the starting 4 configuration, with each face just one colour, you can never get to a configuration where

a) all except two cubelets are in the correct position, or
b) all corner cubelets are in the correct position and all except one are correctly oriented, or
c) all edge cubelets are in the correct position and all except one are correctly oriented.

Prove that it is really impossible to reach any of these configurations (unless you disassemble the cube).

A A survey of problem-solving strategies

This is a survey of the problem-solving strategies that you have learned about in this book.

The main steps in problem-solving are:

1. **Understand the problem**
2. **Investigate the problem**
3. **Write up the solution properly**
4. **Review your solution**

Usually the hardest step is the investigation. But the other steps are just as important: obviously, understanding the problem is essential and will help in the investigation. By writing up the solution properly you communicate it to others and also check that your solution is complete and correct. The review provides another check, and also helps you to transfer your insights to problems that you will solve in the future.

1. Understand the problem

Read the problem carefully.

- ❑ What is given?
- ❑ What are we looking for?
- ❑ What are the premises?

2. Investigate the problem

Suppose you are feeling **'I have no idea what to do!'** What should you do?

Then you take out your **toolbox.** It contains **general strategies, special strategies, techniques,** and **concepts.**

Keep going, be stubborn but also **flexible:** if one strategy does not work then try another. Work **step by step.**

D. Grieser, *Exploring Mathematics*, Springer Undergraduate Mathematics Series, https://doi.org/10.1007/978-3-319-90321-7

General strategies

- ❑ **Get a feel for the problem,** get well acquainted with it: Consider special cases or examples, make sketches and tables
- ❑ Have I seen a **similar problem** before?
- ❑ Is there a **simpler problem** that I could consider first?
- ❑ How can I use what is given? – **Working forward**
- ❑ How can I reach the goal? – **Working backward**
- ❑ Can I formulate useful **interim goals?**
- ❑ Introduce **notation**
- ❑ Formulate **conjectures**

Here are some other questions that you may ask yourself:

- ❑ Is there a **pattern?**
- ❑ What is **essential,** what matters (and what doesn't)?

Special strategies

Special strategies are useful for particular classes of problems.

- ❑ Recursion, induction
- ❑ Counting by bijection, counting in two ways
- ❑ Pigeonhole principle
- ❑ Extremal principle
- ❑ Invariance principle

If you think that one of these principles could be useful, **plan** how to use it: what would I need to do for the inductive step, what could be the pigeons and the holes, which quantity could we maximise/minimise to solve our existence problem, what could be an invariant that helps to prove non-existence?

Techniques, concepts

There are many techniques and concepts that can be useful for solving problems. Some that you learned about in this book are:

- Solving recursions using the ansatz that $a_n = \alpha^n$ and the superposition principle
- Graphs: using them to represent complex webs of relationships; even/odd arguments, EULER's formula
- Counting principles
- Congruences
- Signature of a permutation

There are many more techniques and concepts that you will learn about when studying more mathematics.

Of course with **experience** you will get better at solving problems. You acquire it by **solving problems yourself** and also by **understanding given solutions.** After a while you will build up a **stock of problems and solutions,** which you can use with the strategy "similar problems". Or you may follow a middle course: first think about the problem, then read a little, then try to make progress yourself, then get another idea from the book, etc. This can be an efficient way to build up your stock and gain experience.

Many solutions to problems have solidified into mathematical theories (for instance number theory, or graph theory) which have been refined more and more over the centuries. When you study such a theory you will learn more techniques, concepts and ideas. Always ask yourself: Which problems does the theory help me to solve? Which problems can I solve now that I could not solve before?

And if none of the tools help – build your own. This is the high art of problem-solving and of mathematical research.

3. Write up the solution properly

Be very careful to:

- Argue conclusively, use correct logic

- ❑ Structure sensibly
- ❑ Write understandably (mention ideas and motivations, make a sketch etc.)

Mathematical terminology often helps us to say things precisely. But it should be accompanied by explanations: a sequence of equations alone is usually hard to understand. Words are important!

4. Review your solution

When you have solved the problem you should ask yourself more questions, for example:

- ❑ What have I learned?
- ❑ Is the solution reasonable?
- ❑ Is there a different or even a better way to do it?
- ❑ Did I use all the premises? How did they matter?
- ❑ Can I find interesting problems by modifying the question, the premises?

EUREKA!

B Basics on sets and maps

Modern mathematics is formulated in the language of sets and maps (also called mappings). This language is precise and gives a common basis to the enormous breadth of the subject. Therefore we also use it at many places in this book. Here you will find the essential definitions and facts.

Sets and elements

A **set** is a collection of objects, its **elements,** where it only matters whether an element belongs to the set or not. That is, the same element cannot occur several times in a set, and order does not matter. The elements are listed between braces $\{$ and $\}$.

For example, $\{1,2\}$ and $\{5,8,9\}$ are sets. The sets $\{1,2\}$, $\{2,1\}$ and $\{1,1,2\}$ are all the same. (Switching order and multiple entries are allowed but have no effect.)

If an element belongs to a set then we write $\in$, otherwise $\notin$. For example $1 \in \{1,2\}$ but $3 \notin \{1,2\}$. (In words: 1 is an element of $\{1,2\}$, 3 is not an element of $\{1,2\}$.)

A **subset** of a set A is a set B all of whose elements are also in A. Then we write $B \subset A$. For example, $\{1,2\} \subset \{1,2,5\}$.

Given a set, you can select from it elements having special properties. This is indicated by a colon (sometimes a vertical line $|$ is used).

Example:
Let $A = \{1,2,3,4,5,6\}$. Then $\{n \in A : n \text{ is even}\} = \{2,4,6\}$.
(In words: the set of elements n of A which are even)

Number systems are sets:

$\mathbb{N} = \{1,2,3,\dots\}$	the set of **natural numbers**
$\mathbb{Z} = \{\dots,-2,-1,0,1,2,\dots\}$	the set of **integers**
$\mathbb{Q} = \{\frac{p}{q} : p \in \mathbb{Z},\ q \in \mathbb{N}\}$	the set of **rational numbers**
$\mathbb{R}$	the set of **real numbers**

D. Grieser, *Exploring Mathematics*, Springer Undergraduate Mathematics Series, https://doi.org/10.1007/978-3-319-90321-7

In addition we write $\mathbb{N}_0 = \{0,1,2,\dots\}$. In this book we do not need the precise definition of $\mathbb{R}$.

It is useful to have a symbol for the set that has no elements, called the **empty set**. This set is denoted $\emptyset$. If A is any set then $\emptyset \subset A$.

The elements of a set need not be numbers, they can be any objects, for example graphs. Since sets are objects, they can also be elements of another set. For example, the set $\{\{1,2\},\{1,3,6\}\}$ has the two elements $\{1,2\}$ and $\{1,3,6\}$, and the set $\{\emptyset\}$ has the element $\emptyset$. If A is a set then you can consider all subsets of A as elements of a new set, called the **power set** of A and denoted $\mathcal{P}(A)$. For example

$$\mathcal{P}(\{1,2\}) = \{\emptyset,\{1\},\{2\},\{1,2\}\}.$$

Note that $\{1\}$ is an element of $\mathcal{P}(\{1,2\})$. However, 1 is not an element of $\mathcal{P}(\{1,2\})$, but it is an element of $\{1\}$, and of $\{1,2\}$. This may appear pedantic at first, but it is consistent, and in the long run this consistency is very useful! So the power set $\mathcal{P}(\{1,2\})$ has four elements: $\emptyset, \{1\}, \{2\}$ and $\{1,2\}$.

If A is a set then $|A|$ denotes the number of elements of A. It can be infinite. For example $|\{3,5\}| = 2$, $|\emptyset| = 0$, $|\mathbb{N}| = \infty$ and $|\mathcal{P}(\{1,2\})| = 4$.

Operations with sets

From two sets you can create new sets using various operations: For sets A and B we define:

Name	Symbol	Definition
union of A, B	$A \cup B$	$\{x : x \in A \text{ or } x \in B\}$
intersection of A, B	$A \cap B$	$\{x : x \in A \text{ and } x \in B\}$
difference of A, B	$A \setminus B$	$\{x : x \in A, \text{ but } x \notin B\}$
product of A, B	$A \times B$	$\{(x,y) : x \in A \text{ and } y \in B\}$

(In words: A union B, A intersected with B, A minus B, A times B.) For example, $\{1,2\} \cup \{2,4,6\} = \{1,2,4,6\}$, $\{1,2\} \cap \{2,4,6\} = \{2\}$ and $\{1,2\} \setminus \{2,4,6\} = \{1\}$. For the product (also called product set or cartesian product) see the next section.

Two sets are called **disjoint** if they have no common elements, i.e. if $A \cap B = \emptyset$. If A, B are disjoint then $A \cup B$ is sometimes called their **disjoint union.**

Several sets $A_1, A_2, \ldots, A_n$ are disjoint if $A_1 \cap A_2 \cap \cdots \cap A_n = \emptyset$, and are called **pairwise disjoint** if every two of them are disjoint: $A_i \cap A_j = \emptyset$ for all $i \neq j$. For example, $A_1 = \{1,2\}$, $A_2 = \{2,3\}$ and $A_3 = \{1,3\}$ are disjoint (there is no element in all of them) but not pairwise disjoint.

Tuples

Tuples are another way of grouping several objects (numbers, graphs, sets etc.) into new objects. After sets, they are the next simplest and most useful way to do this.

From two objects a, b you can form the **pair** (a, b). Here order matters, that is, (a, b) and (b, a) are different if $a \neq b$. If the objects a, b are equal then we still consider this as a pair: (a, a). If A and B are sets then $A \times B$ denotes the set of all pairs (a, b) with $a \in A$ and $b \in B$. For $A = B$ we often write A^2 instead of $A \times A$.
For example, $\{1,2\} \times \{1,2,3\} = \{(1,1),(1,2),(1,3),(2,1),(2,2),(2,3)\}$; you can arrange the elements of $A \times B$ in a rectangular array:

	1	2	3
1	(1,1)	(1,2)	(1,3)
2	(2,1)	(2,2)	(2,3)

The rows correspond to the elements of A, the columns to the elements of B. (One could also do it the other way around, but this representation is most common, for example in the context of matrices.)

From three objects a, b, c you can form the **triple** (a, b, c). Again order matters, and multiple entries are allowed. For example, (a, a, b) is a triple that must not be confused with the pair (a, b). If A, B, C are sets then $A \times B \times C$ denotes the set of all triples (a, b, c) with $a \in A$, $b \in B$, $c \in C$. You could arrange these in a cuboid (box). If $A = B = C$ then we often write A^3 instead of $A \times A \times A$.

From four objects we can form **quadruples**, from five **quintuples**, from k objects k**-tuples** $(a_1, \ldots, a_k)$, where $k \in \mathbb{N}$. So a 2-tuple is a pair, a 3-tuple a triple etc. The objects $a_1, \ldots, a_k$ are called the **components** of the tuple.

The notion of a tuple allows us to talk about higher-dimensional space: Real numbers can be thought of as points on the number line. Pairs of real numbers correspond to points in the plane (the two numbers indicating the coordinates with respect to a chosen coordinate system). Triples of real numbers correspond to points in space. Therefore, for any $n \in \mathbb{N}$ we *define* by analogy n-dimensional space as the set of n-tuples of real numbers. This is an important idea, but we do not use it in this book.

Maps

A **map** or **mapping** is a rule that assigns to each element of a set A an element of a set B. We write $f : A \to B$, and $f(a)$ for the element of B assigned to $a \in A$. Here f is a name for the map.

For example the rule $f(n) = 2n$ defines a map $f : \mathbb{N} \to \mathbb{N}$. Each natural number is mapped to its double.

Instead of $f : \mathbb{N} \to \mathbb{N}$, $f(n) = 2n$ we may write $f : \mathbb{N} \to \mathbb{N}$, $n \mapsto 2n$ (notice the two types of arrows). You may change the name of the variable n: the same map is defined by $f : \mathbb{N} \to \mathbb{N}$, $m \mapsto 2m$ or by $f : \mathbb{N} \to \mathbb{N}$, $\gamma \mapsto 2\gamma$.[1] If you don't need the name f then you may write the map simply as $\mathbb{N} \to \mathbb{N}$, $n \mapsto 2n$.

As another example consider $f : \mathbb{R}^2 \to \mathbb{R}^2$, $(x, y) \mapsto (x, -y)$. The map f describes a reflection in the x axis. That is, $f(x, y)$ is the point in the plane that you obtain by reflecting the point (x, y) across the x axis.

A map taking values in numbers (i.e. $B = \mathbb{R}$ or another number system) is called a **function.** A map $f : A \to B$ is called

injective if for each $b \in B$ there is *at most* one $a \in A$ with $f(a) = b$
surjective if for each $b \in B$ there is *at least* one $a \in A$ with $f(a) = b$
bijective if for each $b \in B$ there is *exactly* one $a \in A$ with $f(a) = b$

Put differently, f is injective if it maps any two different elements of A to different elements of B, it is surjective if each element of B occurs as a value $f(a)$, and it is bijective if it is both injective and surjective.

For example the map $f : \mathbb{N} \to \mathbb{N}$, $f(n) = 2n$ is injective but not surjective since for each $b \in \mathbb{N}$ there is at most one n with $2n = b$, but for odd b there is no such n.

[1] However, there are some useful *conventions:* natural numbers are usually denoted n, m, k, l, i, j; real numbers x, y, z. But there is no strict rule about this.

If E denotes the set of even natural numbers then the map $f : \mathbb{N} \to E$, $f(n) = 2n$ is bijective, because for each even number $b \in E$ there is exactly one $n \in \mathbb{N}$ with $2n = b$. This shows that the properties of a map $f : A \to B$ depend on the sets A and B, called the domain and codomain of f respectively.

A bijective map is also called a **bijection.** It sets up a one-to-one correspondence between the elements of A and B, i.e. each element of A corresponds to exactly one element of B and each element of B corresponds to exactly one element of A.

A map $g : B \to A$ is called the **inverse map** of a map $f : A \to B$ if "the rules for f and g cancel each other out". More precisely, for all $a \in A$ we have $g(f(a)) = a$ (first applying f to a and then applying g to the result will give back a), and for all $b \in B$ we have $f(g(b)) = b$.

Since bijective maps are used repeatedly in this book we note:

Theorem Let $f : A \to B$ be a map. Then the following statements are equivalent (i.e. each implies the other):

a) f is bijective

b) f is surjective and injective.

c) There exists an inverse map for f.

Proof.
The equivalence of a) and b) follows directly from the definition. Suppose c) holds, i.e. f has an inverse map $g : B \to A$. Then f is surjective since to any $b \in B$ we can take $a = g(b)$, this satisfies the equation $f(a) = b$ since $f(g(b)) = b$.

f is also injective: if $f(a) = b$ then applying g to both sides yields $g(f(a)) = g(b)$. By assumption $g(f(a)) = a$, so we get $a = g(b)$. We have shown that there can be only *one* $a \in A$ satisfying $f(a) = b$. Thus we have shown that c) implies b) and hence a).
It remains to prove that a) implies c). Assume f is bijective. Then we can write down g: Let $b \in B$. Since f is bijective there is exactly one $a \in A$ with $f(a) = b$. We define $g(b) = a$. Then g is an inverse map of f.
q. e. d.

Examples

Let E be the set of even natural numbers. The inverse map of $f : \mathbb{N} \to E, n \mapsto 2n$ is $g : E \to \mathbb{N}, m \mapsto \frac{m}{2}$.[2]

Let $f : \mathbb{R} \to \mathbb{R}$, $x \mapsto x^3$. This map is bijective with inverse map $g : y \mapsto \sqrt[3]{y}$.

Let $f : \mathbb{R} \to \mathbb{R}$, $x \mapsto x^5 + x$. Is f bijective? This would mean that for each $y \in \mathbb{R}$ there is exactly one $x \in \mathbb{R}$ satisfying $x^5 + x = y$. So we are talking about solving an equation. Using tools from analysis one can show that indeed for each y there is exactly one solution x (compare the discussion on proofs of existence in Section 7.2).[3] So we know that f is bijective. The inverse map is $g : \mathbb{R} \to \mathbb{R}$, $g(y)$ = the unique solution x of the equation $x^5 + x = y$. This defines g, even though we did not write down an expression for $g(y)$. In fact, it can be proved that there is no such explicit expression (for example using roots and arithmetic operations).

The map $f : \{1,2,3\} \to \{1,2,3\}, 1 \mapsto 2, 2 \mapsto 3, 3 \mapsto 1$ is bijective with inverse map $g : \{1,2,3\} \to \{1,2,3\}, 2 \mapsto 1, 3 \mapsto 2, 1 \mapsto 3$.

The examples show the most important ways of defining a map $f : A \to B$: by a formula, by a prescription (g in the third example) or by listing the function values for each argument (if A is finite, as in the last example). They also show that finding the inverse map amounts to solving an equation: in order to find the inverse map of $f : \mathbb{R} \to \mathbb{R}$, $x \mapsto x^3$ we need to solve the equation $y = x^3$ for x. Taking the cube root yields $x = \sqrt[3]{y}$, so $g : y \mapsto \sqrt[3]{y}$ is the inverse map.

[2] It is useful and helps to avoid confusion and mistakes to use a different variable name for the inverse map, for example m instead of n.

[3] This follows from the intermediate value theorem and the following properties of f: the map f is continuous and strictly monotone, and $f(x) \to \infty$ for $x \to \infty$, $f(x) \to -\infty$ for $x \to -\infty$.

List of symbols

$\forall, \exists$	'For all', 'There is'	160
$\neg$	negation: if A is a proposition then $\neg A$ is the proposition that A is false	158
$\Rightarrow, \Leftrightarrow$	implication and equivalence	161
$:=$	is defined as	66
$\subset$	is a subset of	285
$\in$	is an element of	37
$\cup, \cap, \setminus, \times$	the set operations union, intersection, set difference, product set	108, 286
$\emptyset$	the empty set	286
$\lvert X \rvert$	number of elements (cardinality) of the set X	107
$(a,b), (a,b,c)$	a pair or triple	113, 287
$\mathbb{N}, \mathbb{Z}, \mathbb{Q}, \mathbb{R}, \mathbb{N}_0$	natural numbers, integers, rational numbers, real numbers, natural numbers including zero	285
$\rightarrow, \mapsto$	mapping arrows: $f : A \rightarrow B$ denotes a map from a set A to a set B. Instead of $f(a) = b$ one may write $f : a \mapsto b$	288
Σ	short notation for sums. Instead of $f(1) + f(2) + \cdots + f(n)$ one writes $\sum_{k=1}^{n} f(k)$, or $\sum_{k=1}^{n} f(k)$ to save printing space, where f is any expression (i.e. a function of k). For example $\sum_{k=1}^{n} k = 1 + 2 + \cdots + n$ and $\sum_{k=1}^{n} k^2 = 1^2 + 2^2 + \cdots + n^2$	66
$\lfloor x \rfloor$	GAUSS bracket of the real number x: the value of $\lfloor x \rfloor$ is the largest integer which is less than or equal to x	205
$\mathrm{frac}(x)$	The part after the decimal point of a positive real number x; that is, $\mathrm{frac}(x) = x - \lfloor x \rfloor$	205
n	used throughout as notation for an arbitrary natural number	25
$n!$	factorial: $n!$ = the product of the numbers $1, 2, \ldots, n$	25, 115
$\binom{n}{k}$	binomial coefficient: $\binom{n}{k} = \dfrac{n(n-1)\ldots(n-k+1)}{k!}$	116
$\mid$	divides: $n \mid a$ means that n is a divisor of a	181
$\equiv$, mod	congruent, modulo: $a \equiv b \mod n$ means that $b - a$ is divisible by n	188
d_V	degree of the vertex V in a graph (number of edges containing V, with loops counted twice)	93
b_F	number of border edges of a face F of a plane graph	91
$\ell(\overline{AB})$	length of the straight line from A to B	232

D. Grieser, *Exploring Mathematics*, Springer Undergraduate Mathematics Series, https://doi.org/10.1007/978-3-319-90321-7

Glossary

addition rule

Basic counting rule. 108

axiom

A proposition which is taken to be true in a given context. 168

bijection, bijective map

A map $f : A \to B$ is bijective if for every $b \in B$ there is exactly one $a \in A$ satisfying $f(a) = b$. 117, 288

binomial coefficient

The binomial coefficient $\binom{n}{k} = \frac{n(n-1)\ldots(n-k+1)}{k!}$ is the number of k-element subsets of an n-element set. 116

binomial theorem

The formula $(a+b)^n = a^n + \binom{n}{1}a^{n-1}b + \binom{n}{2}a^{n-2}b^2 + \cdots + \binom{n}{n-1}ab^{n-1} + b^n$ for real numbers a, b and $n \in \mathbb{N}_0$. It generalises the well-known formula $(a+b)^2 = a^2 + 2ab + b^2$.. 136

cardinality

The cardinality of a finite set X is the number of elements of X. Notation: $|X|$. Two (possibly infinite) sets are said to have the same cardinality if there is a bijection between them. 107, 118, 133

Catalan numbers

Sequence of numbers which occurs when counting triangulations, for example. 67

colouring problems

An interesting class of problems for graphs. 101

conclusion

The proposition B in an implication $A \Rightarrow B$. 161

congruent

For integers: For $n \in \mathbb{N}$ and $a, b \in \mathbb{Z}$ we say that a, b are congruent modulo n if a and b leave the same remainder when divided by n, or equivalently if $b - a$ is divisible by n. In geometry: Two figures (i.e. subsets of the plane or of space) are congruent if one can be transformed into the other by a rigid motion, i.e. by a translation, rotation or reflection. 188

connected

A set is connected if you can get from any point (i.e. element) of the set to any other point of the set along a path which runs inside the set. Here the set can be a graph or a subset of the plane or of space. 84

D. Grieser, *Exploring Mathematics*, Springer Undergraduate Mathematics Series, https://doi.org/10.1007/978-3-319-90321-7

counting in two ways

Method of deriving identities by counting the elements of a set in two different ways. 123, 169

degree of a vertex

In a graph the degree of a vertex V is the number of edges containing V (loops are counted twice). Notation: d_V. 93, 198

dense

For subsets $A \subset B \subset \mathbb{R}$ we say that A is dense in B if for any two numbers $x, y \in B$ with $x < y$ there is an element $r \in A$ satisfying $x \leq r \leq y$. 211, 246

disjoint

Two sets A, B are called disjoint if they have no common elements, i.e. if $A \cap B = \emptyset$; several sets are called pairwise disjoint if any two of them are disjoint. 286

division with remainder

For $a \in \mathbb{Z}$, $n \in \mathbb{N}$ the representation of a as $a = qn + r$ with $q, r \in \mathbb{Z}$ and $0 \leq r < n$. The remainder is r. 184

divisor

If $a, n \in \mathbb{Z}$ then n is a divisor of a if a is divisible by n, that is if there is $q \in \mathbb{Z}$ satisfying $a = qn$. 181

domino effect

Intuition for mathematical induction. 71

edge-face formula

The formula $2e = \sum_{F \text{ face of } G} b_F$ for plane graphs G, where e is the number of edges and b_F is the number of boundary edges of the face F. 91

edge-vertex formula

The formula $2e = \sum_{V \text{ vertex of } G} d_V$ for graphs, where e is the number of edges and d_V is the degree of the vertex V. 92

(logically) equivalent

Two propositions A, B are (logically) equivalent, in symbols $A \iff B$, if both 'A implies B' and 'B implies A' are true. For example, the propositions $n + 3 = 8$ and $n = 5$ about natural numbers n are equivalent. 163, 164

EULER's formula

The formula $v - e + f = 2$ for planar graphs or for polyhedra, where v is the number of vertices, e is the number of edges and f is the number of faces. 84, 98

extremal principle

Important idea for proofs of existence. Fundamental principle of science that extremal configurations have special significance. 173, 217

face of a plane graph

One of the regions into which a plane graph subdivides the plane. 84

factorial

n factorial $= n! =$ the product of the numbers $1, 2, \ldots, n$, e.g. $4! = 1 \cdot 2 \cdot 3 \cdot 4 = 24$. 115

FERMAT numbers

The numbers $F_n = 2^{(2^n)} + 1$ where $n \in \mathbb{N}_0$. 166, 168, 173

FIBONACCI numbers

The sequence of numbers $1, 1, 2, 3, 5, 8, \ldots$ defined by $a_0 = a_1 = 1$ and $a_n = a_{n-1} + a_{n-2}$ for $n \geq 2$. 54

GAUSS trick

Method for calculating $1 + \cdots + n$ quickly. 33, 124

graph

A finite structure given by a set of objects, called vertices, and a set of edges, each joining two vertices. Examples: 1. a plane graph; 2. vertices are the people in a room, and two people are joined by an edge if they know each other. 93

graph, plane

A graph which is drawn in the plane in such a way that vertices are points, edges are lines (not necessarily straight) joining the vertices, and the lines don't intersect (except at vertices). 83

implication

Proposition of the form 'A implies B' ($A \Rightarrow B$). 161

in general position

A set of lines is in general position if no three go through a point and no two are parallel. 28

infinite descent

Method which is sometimes useful for proving that an equation has no solution in natural numbers: you prove that if there was a solution then there would have to be another solution which is smaller. 235

injective, one-to-one, 1-1

A map $f : A \to B$ is injective (or one-to-one) if for every $b \in B$ there is at most one $a \in A$ satisfying $f(a) = b$. 288

invariance principle

Important idea for proofs of nonexistence. Fundamental principle of science that in complex processes the quantities that don't change have special significance. 176, 251

invariant

Something that does not change during a certain process. 256

inverse map

$g : B \to A$ is called the inverse map of the map $f : A \to B$ if $g(f(a)) = a$ and $f(g(b)) = b$ hold for all $a \in A$, $b \in B$. A map f has an inverse map if and only f is bijective. 289

irrational number

A number which cannot be written as the quotient of two integers. 175, 203

lattice point

A point of the plane both of whose coordinates are integers. 215

lemma

An auxiliary statement which is used in the proof of a theorem. 182

loop in a graph

An edge of a graph whose two endpoints coincide. 93

map, mapping

A prescription which assigns to each element of a set A an element of a set B. We write $f : A \to B$ where f is a name for the map. 288

mean: harmonic, geometric, arithmetic or quadratic

For $a, b > 0$ we call $\frac{2}{\frac{1}{a}+\frac{1}{b}}$ the harmonic mean, $\sqrt{ab}$ the geometric mean, $\frac{a+b}{2}$ the arithmetic mean, and $\sqrt{\frac{a^2+b^2}{2}}$ the quadratic mean of a and b. 221, 234, 246

modulo, mod

The remainder of a modulo n is the remainder when dividing a by n. 184

multiple

If $a, n \in \mathbb{Z}$ then a is a multiple n if a is divisible by n, that is if there is $q \in \mathbb{Z}$ satisfying $a = qn$. 181

multiple counting principle

A useful idea for many counting problems. 110

multiplication rule

Basic counting rule. 108

necessary

Proposition A is necessary for proposition B if B implies A. 163

ordering

An order in which you can line up some objects, for example the numbers $1, \ldots, n$. 115, 262

parity

Property of an integer being even or odd. 150, 254

PASCAL's triangle

Arrangement of the binomial coefficients in a triangular array. 79, 120, 127

path in a graph

A sequence of edges in a graph where each edge has a common vertex with the previous edge, so that the edges can be traversed one after another. The path is called closed if the first and last vertex are the same. 84

permutation

A way to order the elements of a set M, e.g. $M = \{1, \ldots, n\}$; may also be interpreted as bijection $M \to M$. 115, 262

pigeonhole principle

Fundamental idea for proofs of existence: If $n+1$ pigeons sit in n pigeonholes then at least one hole must be occupied by more than one pigeon. 173, 195

polygon

A plane figure bounded by finitely many straight lines, for example a triangle or hexagon. A polygon is called **convex** if the straight line connecting any two of its vertices lies in the polygon. 61

power set $\mathcal{P}(A)$

Set of subsets of the set A. 44, 118, 286

predicate

A statement containing variables, which becomes a proposition when values are plugged in for the variables. 158

premise

The proposition A in an implication $A \Rightarrow B$. 161

prime factorisation

The representation of a natural number $n > 1$ as product of prime numbers. 183

prime number

A natural number which has precisely two divisors. 182

proof by contradiction

In order to show $A \Rightarrow B$ you prove that 'A and $\neg B$' implies a false statement (a contradiction). 163, 167

proof by mathematical induction

A type of proof which is often useful for proving statements about all natural numbers. 71, 169

proof of existence

A proof which shows the existence of an object with certain properties. 169, 256

proof of impossiblity

Proof that something is impossible. 98, 174, 252, 272

proof of nonexistence

A proof which shows that an object with certain properties cannot exist. 174, 257

proof without words

Sketch or diagram which suggests a proof of an arithmetic or geometric proposition. 128

proof, direct

In order to show $A \Rightarrow B$ you prove a sequence of implications $A \Rightarrow A_1 \Rightarrow A_2 \Rightarrow \cdots \Rightarrow B$. 167

proof, indirect

In order to show $A \Rightarrow B$ you prove $\neg B \Rightarrow \neg A$. 163, 167

proposition

A statement which can be true or false. 157

rational number

A number that can be written as the quotient of two integers. 201, 285

recurrence relation

An equation giving each term in a sequence as a function of previous terms. It expresses the idea of recursion in a counting problem. 41

recursion

A problem solving technique: reduce the problem to a smaller problem of the same kind. 141

remainder

The remainder of the division of $a \in \mathbb{Z}$ by $n \in \mathbb{N}$ is the number r in the representation $a = qn + r$ where $q, r \in \mathbb{Z}$ und $0 \leq r < n$. 184, 199

semi-invariant

Something that changes in a controlled way during a certain process. 264

shift by one

Typical phenomenon in counting problems, frequent source of errors. For example, a train with 10 cars has 9 couplings, and the number of integers from 10 to 100 is 91, not $100 - 10 = 90$, since one needs to count both the first and last. 23, 109, 149

signature of a permutation

The parity of the number of inversions of a permutation. 262

sufficient

Proposition A is sufficient for proposition B if A implies B. 163

superposition principle

In linear problems (equations) one may obtain new solutions by adding two solutions or multiplying a solution by a constant. 57

surjective, onto

A map $f : A \to B$ is surjective (or onto) if for every $b \in B$ there is at least one $a \in A$ satisfying $f(a) = b$. 288

to tile

To cover completely without overlaps, and without pieces (tiles) extending over the boundary. 50, 253

topology

Area of mathematics which studies those properties of shapes (or other mathematical structures, like maps) which remain the same when the shapes (or maps etc.) are changed in a continuous way. 99

torus

A surface which looks essentially like a bike tyre inner tube (without valve). 99

trapezoidal number

Natural number that can be represented as the sum of several consecutive natural numbers. 146

triangle inequality

In a triangle each side is shorter than the sum of the lengths of the two other sides. 221

triangulation

subdivision of a polygon into triangles by non-intersecting diagonals. 60

truth value

true (**t**) or false (**f**). 158

Lists of problems, theorems and methods

List of problems

D. Grieser, *Exploring Mathematics*, Springer Undergraduate Mathematics Series, https://doi.org/10.1007/978-3-319-90321-7

List of theorems and methods

Hints for selected exercises

Exercises in Chapter 1

E 1.5 First consider the powers of two: 4, 8, 16 etc. The numbers in between can be dealt with similarly to the next larger power of two. To argue that your answer is best possible consider this question: how much can the number of pieces grow with each cut? The smallest number of cuts is $\lceil \log_2 n \rceil$ where $\log_2$ is the logarithm to base 2 and $\lceil x \rceil$ is x rounded up to the next integer.

E 1.6 Here the Gauss bracket is useful, the symbol for rounding down to the previous integer. By definition $\lfloor x \rfloor =$ the largest integer which is $\leq x$. For example $\lfloor \frac{7}{2} \rfloor = 3$ is the number of even natural numbers which are less than or equal to 7.

E 1.7 To recognize the patterns it is useful to know the values of $n!$ for $n = 1,\ldots,5$ by heart. To prove the formulas which you might conjecture from this the **telescope trick** is useful: for example in a) write $1 \cdot 1! = (2-1)1! = 2! - 1!$, $2 \cdot 2! = (3-1)2! = 3! - 2!$, and in general $n \cdot n! = \cdots = (n+1)! - n!$. Then

$$\begin{aligned} 1 \cdot 1! + 2 \cdot 2! + \cdots + n \cdot n! &= (2! - 1!) + (3! - 2!) + \cdots + ((n+1)! - n!) \\ &= (n+1)! - 1! = (n+1)! - 1, \end{aligned}$$

since 2! cancels with $-2!$, similarly 3! with $-3!$ (in the next term), up to $n!$ with $-n!$.

The telescope trick also helps with b) and c).

The sum in c) is $n + 1 - \frac{1}{n+1}$.

E 1.8 a) $a_n = n + 2$ b) $a_n = \frac{3}{2} \cdot (n+1)!$ c) $(-1)^{n-1}$
d) $a_n = a_{n-1} + a_{n-2} + a_{n-3}$ where $a_1 = a_2 = a_3 = 1$
e) $a_n = n2^n$

E 1.9 For example, you could try to prove the recursion $s_n = 2s_{n-1}$.

D. Grieser, *Exploring Mathematics*, Springer Undergraduate Mathematics Series, https://doi.org/10.1007/978-3-319-90321-7

E 1.10 Consider $n = 1, 2, \ldots$, but also $n = 40$. Can you do this without a calculator?

E 1.11 Consider the number of lines in each group of parallels. For example, if the lines ℓ_1, ℓ_2 are parallel and the lines ℓ_3, ℓ_4, ℓ_5 are parallel to each other but not to ℓ_1, ℓ_2 then these numbers are 2 and 3. If no two lines are parallel then these numbers are all equal to 1.

Exercises in Chapter 2

E 2.1 Use $|\gamma| < 1$.

E 2.4 For $a_n = 2a_{n-1} - a_{n-2}$ a second solution is $a_n = n$. In general: $n\alpha^n$.

E 2.10 Distinguish two kinds of subsets: those that contain the element n and those that don't.

E 2.12 When searching for a recursion you may find it difficult to reduce the problem to a smaller problem of the same kind. One way out is to consider a second problem in parallel: Let b_n be the number of tilings of a $3 \times n$ rectangle from which the squares at positions $(n, 2)$ and $(n, 3)$ are deleted. Find formulas that express a_n and b_n in terms of $a_{n-1}, b_{n-1}, a_{n-2}, b_{n-2}$. Eliminate b_n.

E 2.13 See the hint for Exercise E 2.12.

E 2.14 First check for $n = 1, 2, 3, 4, 5$ whether the first player can force victory. How can you reduce the answer for a game with n matches to the answer for a smaller number of matches?

E 2.15 The table suggests the Fibonacci recursion $o_n = o_{n-1} + o_{n-2}$. Directly from the problem you can deduce the recursion $o_n = o_{n-1} + o_{n-3} + o_{n-5} + \ldots$ (where we set $o_i = 0$ if $i \leq 0$). How? How can you derive the Fibonacci recursion from this?

Exercises in Chapter 3

E 3.1 Prove by induction that for each $n \in \mathbb{N}$ you can put n pins into the suitcase :-).

E 3.4 What could be the inductive step? First try to understand the step from $n = 2$ to $n = 3$.

E 3.7 Compare Exercise E 2.10.

E 3.9 The main task is to show that this rule makes sense: there could be several paths from A to B, and you need to show that each path yields the same resulting colour for B. Here is an idea how to prove this: if you have two such paths then you can deform or 'slide' one path so it becomes the other path. What happens to the number of intersection points when you slide across a vertex in the process?

Exercises in Chapter 4

E 4.6 What can you say about the numbers b_F in a hypothetical drawing without intersections?

E 4.7 Use the graph formulas.

E 4.8 Cut the torus open so you can flatten it into the plane and consider the resulting plane graph.

E 4.9 Your argument could begin as follows: the points D, E must either both lie in the interior region or both lie in the exterior region. Suppose they lie in the interior region (the other case is similar). Now consider the connections from D to A, B, C and think about where E could lie.

E 4.10 Have you seen a similar problem before?

E 4.11 Use the graph formulas.

E 4.12 Try induction on f. Let F be a face with at most 5 borders. How can you remove this face and produce a smaller graph, so that any proper 6-colouring of the smaller graph can be modified into a proper 6-colouring of the original graph?

E 4.13 Use the representation of the torus as a square with opposite sides glued together as in Figure 4.12.

E 4.14 Proceed similarly to the solution of Exercise E 4.12.

E 4.15 Pay attention to the vertex degrees. Why can't you start at the top of the roof?

E 4.16 Use the graph formulas to derive restrictions on the possible numbers of vertices per face and edges per vertex.

E 4.17 Choose a point in the interior of each small triangle. Construct a graph as follows: the vertices are these chosen points. Connect two vertices by an edge if the corresponding small triangles have an edge in common and the labels at the endpoints of this edge are different.

Exercises in Chapter 5

E 5.4 Number of hands: $\binom{52}{5}$.
One pair: choose the rank of the pair, then the two suits of the pair, then the ranks of the three other cards, then their suits. This results in $13 \cdot \binom{4}{2} \cdot \binom{12}{3} \cdot 4^3$ hands.
Therefore, about 42% of all hands have one pair, so the probability of getting one pair is about 0.42.

E 5.6 First count: find the number of quadruples for each fixed d. Second count: find the number of quadruples in each of these three cases: a, b, c are all distinct; two of the three numbers a, b, c are equal; and $a = b = c$.

E 5.11 Analogous to Exercise E 5.10.

E 5.13 For n odd we have $e_n = o_n$, for n even we have $e_n = o_n + 1$. Bijection proof for n odd: if $a < n$ then map (a, b) to $(a + 1, b)$, otherwise to $(0, b)$.

A possible generalisation: Let $k \in \mathbb{N}$ and $n_1, \ldots, n_k \in \mathbb{N}_0$. Consider k-tuples $(a_1, \ldots, a_k)$ with $a_i \in \{0, 1, \ldots, n_i\}$ for each i. Let e be the number of such k-tuples with $a_1 + \cdots + a_k$ even and o the number with $a_1 + \cdots + a_k$ odd. Then $e = o + 1$ if all n_i are even, and $e = o$ otherwise.

E 5.14 Distinguish two cases: Either the subset contains the element $n + 1$ or it does not.

E 5.15 When multiplying out $(a + b)^n = (a + b)(a + b) \cdots (a + b)$, in which ways can the term $a^{n-k}b^k$ arise?

E 5.16 b) $\binom{n+1}{1} + 5\binom{n+1}{3} + 5^2\binom{n+1}{5} + \ldots$ is divisible by 2^n. Question for further investigation: Can you find a direct proof for the formula $2^n a_n = \binom{n+1}{1} + 5\binom{n+1}{3} + 5^2\binom{n+1}{5} + \ldots$? For example, by counting a certain set in two ways?[1]

E 5.17 Add $(1-\sqrt{2})^n$ to a^n. Use the idea of Exercise E 5.16 a). How big is $1-\sqrt{2}$ approximately?

E 5.18 You could use the identity of Exercise E 5.14.

E 5.19 Imagine n balls lying in a row. Decompositions of n corresponds to decompositions of the set of balls into connected groups. In what other way could you fix such a decomposition into groups? If needed, draw all decompositions for $n = 3$.

E 5.20 Find a bijection from the set $\{(a,b,c) \in \mathbb{N}^3 : a+b+c = n\}$ to the set $\{(x,y) \in \mathbb{N}^2 : x < y < n\}$.

E 5.23 The identity is $\binom{n}{3} = \binom{n-1}{2} + \binom{n-2}{2} + \cdots + \binom{2}{2}$ for $n \geq 3$.

E 5.24 For $n = 2,3,4,5$ the numbers are $2,4,8,16$, so you might conjecture the formula 2^{n-1}. However, for $n = 6$ you only get 31 regions. To find a general formula it may be helpful to look at Exercise E 5.7.

Exercises in Chapter 6

E 6.3 Problem and solution formula are symmetric; the derivation is not as is clear from the solution scheme in Figure 6.9. This asymmetry has the effect that there are different paths to the result: instead of starting with the bottom rectangle (the a,b rectangle) and its diagonal we could have argued starting with one of the other boundary rectangles (a,c rectangle or b,c rectangle).

E 6.4 One idea: delete a suitable part of $14 = (-1) + 0 + 1 + 2 + 3 + 4 + 5$.

Another idea: for example $n = 14 = 2 \cdot 7$ could be written using $7 = 3 + 4$ as $14 = (3+4) + (3+4)$, then decrease one 3 and increase one 4 by one to get $14 = 2 + 3 + 4 + 5$.

[1] I would be interested to hear about it if you find one.

E 6.5 Build on the solution of Problem 6.3. It is useful to represent n as a product of prime powers (prime factorisation, discussed in detail in Chapter 8).

E 6.6 Use the formula $(a+b)(a-b) = a^2 - b^2$.

E 6.7 Work backward. Write 11111 as a product of prime numbers.

E 6.8 Work backward. The goal can be formulated as $a.a_1 \dots a_n < \sqrt{N} < a.a_1 \dots a_n + 10^{-n}$ where $a \in \mathbb{N}$.

E 6.9 Work forward. How can I use the data? For example you could connect B, C. There is more than one solution.

E 6.10 What could be the last step before such a configuration is reached?

E 6.11 What does it mean for a number n that the card at position n is face down at the end?

E 6.12 Either write down examples and guess ('see') a formula (and then prove it!), or else work systematically: number the customers $1, \dots, n$ where 1 is the last customer etc. Let x_i be the number of apples left after customer $i+1$ has gone, then $x_0 = 1$ and $x_i = 2x_{i-1} + 2$ for $i = 1, \dots, n$. A clever way to solve this recursion is to add 2 on both sides and to write the result as $x_i + 2 = 2(x_{i-1} + 2)$.

E 6.13 Work backward. Write $x \overset{?}{>} 0$ and try to eliminate the roots (add $\sqrt{2}$, square etc.) Be careful with signs.

Exercises in Chapter 7

E 7.3 b), d), e) and f) only.

E 7.4 All propositions imply b). There are no other implications. To show this give a counterexample for each of the other implications.

E 7.6 b) says that the set S_1 is infinite.

E 7.7 If n is not a prime number then it has a divisor which is bigger than 1 and at most $\sqrt{n}$. This is true (and useful if you want to check in your head whether 113 is a prime number). Negation (which is false):

$$\exists n \in \mathbb{N} : (\exists m \in \mathbb{N} : 1 < m < n \text{ and } m|n) \text{ and } (\forall m \in \mathbb{N} : m = 1 \text{ or } m^2 > n \text{ or } \neg(m|n))$$

E 7.8 For example: for each person there is a fact which she knows and for which there is no other person knowing it.

E 7.14 First find a necessary condition, then prove that it is sufficient also.

E 7.15 If $\sqrt{2}$ were rational, what could you say about the denominators of the numbers $(\sqrt{2}-1)^n$? Multiply out and collect terms. What happens for large n?

Exercises in Chapter 8

E 8.3 Use $4 \equiv -3 \mod 7$.

E 8.4 You need to show that there are no natural numbers p, q satisfying $2^q = 10^p$.

E 8.6 If n has the decimal representation $a_m \dots a_1 a_0$ then $n = 10^m a_m + \dots + 10a_1 + a_0$. To get a feeling for this, first consider the case $m = 1$, i.e. two-digit numbers.

E 8.7 Use $10a \equiv -a \mod 11$.

E 8.8 Try a similar procedure as for 7 in Exercise E 8.6.

E 8.10 e) Investigate this modulo 4.

E 8.11 Use the uniqueness of prime factorisation.

E 8.12 b) For example in the case $p = 3$: If $3|2b$ then using $3|3b$ we would get $3|(3b-2b)$, hence $3|b$. Try to generalise this. For example for $p = 5$: How can you conclude from $p|2b$ that $p|b$, using $p|5b$? How can you then argue for $p|3b$? When you have understood this, try induction on a for the general case.

c) divide a by p with remainder in order to reduce to b).

Remark on Exercise E 8.12: in most books on number theory these facts are proved using the EUCLIDean algorithm. The proof suggested here is attributed to GAUSS.

E 8.13 The claim is not negated correctly. Instead of "$6^k \not\equiv 6 \pmod{10}$ for all $k \in \mathbb{N}$" the correct negation is "There is $k \in \mathbb{N}$ such that $6^k \not\equiv 6 \pmod{10}$".

For a correct proof use induction.

E 8.15 Find a suitable bijection from the set of divisors to itself.

E 8.16 Write $\lceil k\frac{m}{n} \rceil$ using the remainder r_k of km modulo n. What can you say about $r_1, \ldots, r_{n-1}$?

Result: $\frac{(n+1)(m+1)}{2} - 1$.

Exercises in Chapter 9

E 9.4 Of course you could argue geometrically (enter the triangle, exit the triangle). But you could also use the pigeonhole principle: take the vertices as pigeons. What are the holes?

E 9.5 Consider the vertices of the triangle.

E 9.8 Consider $b_k := a_1 + \cdots + a_k$ for $k \in \{1, \ldots, n\}$.

E 9.10 First solve Exercise E 9.9 and use its result.

E 9.11 Use the result of Problem 9.4.

E 9.12 Let (a, b) and (c, d) be lattice points. What does it mean for a, b, c, d that the midpoint between (a, b) and (c, d) is a lattice point?

E 9.13 Consider the remainders of $a, 2a, \ldots, (b-1)a$ modulo b and use Exercise E 8.11.

E 9.14 Consider the pairs (remainder of f_n modulo 1000, remainder of f_{n-1} modulo 1000).

E 9.15 Think about what it means that there are m, k satisfying $999999 \cdot 10^{k-6} \leq 2^m < 10^k$. Take the logarithm to base 10 and relate this to the problem of the goblin falling into the gap, see Figure 9.3.

E 9.16 The prime factor 2 matters.

E 9.17 First investigate the same problem with 11 replaced by 2 or 3. How many numbers do you need to write down so that the analogous statement is true? Do you see a pattern?

E 9.18 Either imitate the solution of Problem 9.4 (that's a good exercise), or use the result of Problem 9.4.

Exercises in Chapter 10

E 10.2 Is there a largest value for the sum?

E 10.5 This is possible, but a little more complicated than the given solution.

E 10.7 Consider multiples of $\frac{1}{n}$ for sufficiently large n.

E 10.9 What extremal property does a boundary triangle have? The vertices lie 'as close together as possible'. Quantify this.

E 10.13 First find conditions modulo 2 and modulo 3 that a minimal solution must satisfy. Then try to find a solution.

E 10.14 The numbers must all be equal. In a), b) this can be proved using the extremal principle, but in c) you need to find a different argument since a set of positive real numbers need not have a smallest element. (Of course c) implies a).)

Remark: The same conclusion also holds if you write positive real numbers at lattice points in the plane, but this is harder to prove. Try it!

E 10.15 For each colouring consider the number of edges whose endpoints have different colours.

E 10.16 For each splitting into two groups G_1, G_2 consider the number of pairs of people in G_1 who know each other, plus the corresponding number for G_2.

E 10.18 First imagine you started with enough petrol to go around the track once. Start anywhere, take all the petrol from each barrel you pass and note at each point in time how much petrol you have. Consider the minimum.

E 10.19 First show: if person a is the left-hand neighbour of person b and a doesn't know b, then there are persons c and d such that c is the left-hand neighbour of d, a knows c, and b knows d.

Then consider the seating with the least number of 'mistakes'.

E 10.20 One approach works as follows: 1. Prove this for $n = 4$ by setting $b_1 = \frac{a_1+a_2}{2}$, $b_2 = \frac{a_3+a_4}{2}$ and then using Theorem 10.2 several times. 2. Prove it for $n = 3$ and numbers a_1, a_2, a_3 by applying the $n = 4$ case to a_1, a_2, a_3 and $a_4 = \frac{a_1+a_2+a_3}{3}$. 3. Prove in general that knowing the assertion for n implies its validity for $2n$ and for $n - 1$. Conclude using induction.

E 10.22 Set $a_i = x_i - x_{i-1}$ and use the inequality of Exercise E 10.20.

E 10.23 Via calculation: Use the inequality of the geometric and arithmetic mean and the HERON formula: area = $\sqrt{s(s-a)(s-b)(s-c)}$ where $s = \frac{a+b+c}{2}$ and a, b, c are the side lengths of the triangle.

Via geometry: Use the solution of Problem 10.3 to prove the equivalent assertion that among all triangles of given area the equilateral has the smallest perimeter.

E 10.24 Let S be the set of given points. Consider all pairs (ℓ, P) where ℓ is a line containing at least two points of S and P is a point of S not lying on ℓ. Prove that there is at least one such pair. For each such pair consider the distance of P from ℓ.

E 10.25 Choose the n points in such a way that the n-gon they form has maximal perimeter.

Exercises in Chapter 11

E 11.1 No. Why not?

E 11.2 Yes, but this is unnecessarily complicated.

E 11.3 No. The number of white and black squares covered by a domino depends on the position of the domino. With the chessboard colouring it does not depend on the position, and this is essential for the argument.

E 11.5 No, for example this number does not change in the move $1324 \to 1342$.

E 11.6 Use the lemma stated before the theorem.

E 11.8 You always need the same number of steps, no matter how you go about it. Why? What changes in a controlled way at each step?

E 11.9 Read off the numbers in a different order, for example in a wavy line.

E 11.10 Consider a suitable colouring with 3 colours.

E 11.11 Use a suitable colouring with the 'colours' 1,2,3,4.

E 11.13 Think about how to quantify the difference between the initial and the final position.

E 11.14 The armchair's position is described by its place and its orientation (how much it is turned relative to the initial position). How do these interact?

E 11.15 The moves might remind you of adding or subtracting 11. So the divisibility rule for 11 could give you an idea, compare Exercise E 8.7.

E 11.16 Parities help.

E 11.18 The number is always even. What happens when you move p across a diagonal?

E 11.19 The remarks after Problem 11.8 describe problems which might arise. There are no others. Formulate this as a condition on r, g, b and prove that the condition is sufficient, by showing how to get from one position satisfying it (and having at least two balls) to another position satisfying it. Why is this sufficient for the proof?

E 11.20 Play some games with $g = 3, r = 4$ and with $g = 3, r = 3$ to get an idea what happens. Starting with the numbers (r, g), which numbers can result after a move? What remains constant in each move?

E 11.22 Look at some examples. Starting with the numbers (r, g, b), which numbers can result after a move? What remains constant in each move?

E 11.23 Colour the holes suitably with three colours and use the result of Problem 11.8.[2]

E 11.24 You can only get to the fourth row, so the biggest possible m is 4. Why? It is easy to find a sequence of moves which gets you to the fourth row. It is harder to prove that $m = 5$ is impossible. Here is an idea: Suppose you could get a peg to position $(0, 5)$. Now assign numbers to all lattice points in such a way that the sum of numbers at occupied points does not increase during any move which might be done at any time during a game. If you choose the numbers well then the sum of numbers of pegs involved in a putative solution is smaller than the number at position $(0, 5)$.

How do you choose the numbers for this to work? Here the equation $x^2 + x = 1$ will play a role. Also, the geometric series will be useful: if $|x| < 1$ then $1 + x + x^2 + \cdots = \frac{1}{1-x}$.

E 11.25 Find quantities which remain constant when rotating any face by 90 degrees. For example, in a) consider the permutations of the cubelets and show that the parity remains constant.

[2]The statement a) is proved in (A. Bialostocki, An Application of Elementary Group Theory to Central Solitaire, The College Mathematics Journal, v 29, n 3, Mai 1998, 208-212) with the help of group theory. The proof proposed here is simpler.

References

General literature

Engel, A. (1998). *Problem-solving strategies.* Problem Books in Mathematics. Springer, New York, NY.
Extensive collection of mathematical problems at all levels of difficulty, with solutions. Also includes short introductions to problem-solving strategies, from the veteran coach of the German team of the International Mathematical Olympiad.

Zeitz, P. (1999). *The art and craft of problem solving.* Wiley, New York, NY.
Very well written textbook on problem-solving.

Tao, T. C. S. (2006). *Solving mathematical problems. A personal perspective.* Oxford University Press.
Practical guide to problem-solving, with many problems, some of them quite difficult, written by one of the most successful mathematicians of our time.

Mason, J., Burton, L. and Stacey, K. (2010). *Thinking mathematically.* Pearson, 2nd edition.
Practical guide to problem-solving, very detailed, with problems of low to moderate difficulty.

Pólya, G. (2014). *How to solve it. A new aspect of mathematical method.* Princeton University Press
THE classic on problem-solving, first published in 1945, written by one of the grand masters. Also includes theoretical considerations on problem-solving from the perspective of teachers. Worth reading!

Pólya, G. (1981). *Mathematical discovery. On understanding, learning, and teaching problem solving.* John Wiley & Sons.
Similar to (Pólya, 2014). Some examples are discussed at great length.

D. Grieser, *Exploring Mathematics*, Springer Undergraduate Mathematics Series, https://doi.org/10.1007/978-3-319-90321-7

Aigner, M. (2007). *Discrete Mathematics.* American Mathematical Society.
A comprehensive introduction to discrete mathematics, for example counting problems, graph theory and much more.

Aigner, M. and Ziegler, G. M. (2014). *Proofs from the book.* Springer, 5th edition.
A collection of beautiful mathematical proofs.

Hathout, D. (2013). *Wearing Gauss's jersey.* A K Peters/CRC Press.
Nice collection of problems from algebra, combinatorics, trigonometry and number theory, with solutions, which usually involve a clever idea. Some mathematical tools are developed along the way.

Berlekamp, E.R., Conway, J.H. and Guy, R.K. (2001-2004). *Winning Ways for your Mathematical Plays, vol. 1-4.* A K Peters, 2nd edition.
Extensive collection of mathematical investigations of games, including nim, peg solitaire, Rubik's cube and many more.

For Chapter 2

Stanley, R. (2015). *Catalan numbers.* Cambridge University Press.
214 ways in which the Catalan numbers occur; exercises with solutions.

For Chapter 4

Aigner, M. (1987). *Graph theory. A development from the 4-color problem.* BCS Associates.
A general introduction to graph theory, starting with colouring problems, with many historical remarks.

Lakatos, I. (1976). *Proofs and refutations. The logic of mathematical discovery. Edited by John Worrall and Elie Zahar.* Cambridge University Press.
Socratic dialogues on the topic of the title, discussed in the context of the EULER characteristic.

Prasolov, V. V. (1995) *Intuitive topology.* American Mathematical Society.
Very good introduction to the ideas of topology, with many pictures.

Jordan, D. and Huggett, S. (2009). *A topological aperitif.* Springer.
Elementary yet rigorous introduction to the basic ideas of topology, with many pictures and many examples worked out in detail.

Flapan, E. (2016). *Knots, Molecules, and the Universe: An Introduction to Topology.* American Mathematical Society.
Elementary introduction to topology, emphasizing intuition, with applications.

Armstrong, M. A. (1983). *Basic topology.* John Wiley and Sons.
A rigorous introduction to topology.

Conway, J.H., Burgiel, H. and Goodman-Strauss, H. (2008). *The Symmetries of Things.* Taylor & Francis.
Many beautiful pictures and good explanations on the topic of symmetry. Includes a chapter on the Euler characterstic.

For Chapter 5

Pugh, C.C. (2010). *Real Mathematical Analysis.* Springer, 2nd edition.
Introduction to analysis in one and several variables, rigorous and with many good illustrations.

Apostol, T.M. (1967). *Calculus. Vol. I and Vol. II.* Blaisdell Publishing Co., 2nd edition.
A comprehensive introduction to one-variable (vol. I) and multi-variable (vol. 2) calculus, rigorous, with applications and many examples. A classic.

For Chapter 8

Hardy, G.H. and Wright, E.M. (2008). *An Introduction to the Theory of Numbers.* Oxford University Press, 6th edition.
THE classic text on elementary number theory.

Silverman, J. (2012). *A Friendly Introduction to Number Theory.* Prentice-Hall, 4th edition.
Very well written introduction to many aspects of elementary number theory.

For Chapter 10

Nahin, P. J. (2004). *When least is best. How mathematicians discovered many clever ways to make things as small (or as large) as possible.* Princeton University Press, Princeton, NJ.
A diverse collection on the topic of the extremal principle, presented in an entertaining way. Historical account interspersed with modern examples.

Hildebrandt, S. and Tromba, A. T. (1996). *The parsimonious universe. Shape and form in the natural world.* Copernicus, New York.
A diverse collection of geometric extremal problems, with many illustrations and including history and applications.

For Chapter 11

Adams, Colin C. (2004). *The knot book. An elementary introduction to the mathematical theory of knots.* Providence, RI: American Mathematical Society.
Informal and accessible introduction to knot theory, with applications in physics, chemistry and biology.

Printed in the United States
By Bookmasters